OCS Study
MMS 2005-016

Workshop on Socioeconomic Research Issues for the Gulf of Mexico OCS Region

February 2004

Editors

Melanie McKay
Copy Editor

Judith Nides
Production Editor

Prepared under MMS Contract
1435-00-01-CA-31060
by
University of New Orleans
Office of Conference Services
New Orleans, Louisiana 70814

U.S. Department of the Interior
Minerals Management Service
Gulf of Mexico OCS Region

New Orleans
March 2005

DISCLAIMER

This report was prepared under contract between the Minerals Management Service (MMS) and the University of New Orleans, Office of Conference Services. This report has been technically reviewed by the MMS and approved for publication. Approval does not signify that contents necessarily reflect the views and policies of the Service, nor does mention of trade names or commercial products constitute endorsement or recommendation for use. It is, however, exempt from review and compliance with MMS editorial standards.

REPORT AVAILABILITY

Extra copies of this report may be obtained from the Public Information Office (Mail Stop 5034) at the following address:

> U.S. Department of the Interior
> Minerals Management Service
> Gulf of Mexico OCS Region
> Public Information Office (MS 5034)
> 1201 Elmwood Park Boulevard
> New Orleans, Louisiana 70123-2394
> Telephone Numbers: (504) 736-2519
>   1-800-200-GULF

CITATION

This study should be cited as:

McKay, M. and J. Nides, eds. 2005. Workshop on Socioeconomic Research Issues for the Gulf of Mexico OCS Region, February 2004. U.S. Dept. of the Interior, Minerals Management Service, Gulf of Mexico OCS Region, New Orleans, LA. OCS Study MMS 2005-016. 153 pp.

TABLE OF CONTENTS

LIST OF FIGURES

LIST OF TABLES

ACKNOWLEDGMENTS

The Minerals Management Service (MMS) thanks all participants in this workshop. Their active participation and enthusiasm were crucial in the success of this endeavor. In addition we would like to thank the following people who were unable to attend for their assistance in the preparation for the workshop: Dr. Shana Walton, Associate Director at the Deep South Regional Humanities Center, Tulane University; Mr. Joey Brackner M.A., Folklorist for the Alabama State Council on the Arts; and Dr. Tina Bucuvalas, Director of the Florida Folklife Program, Division of Historical Resources, Bureau of Historical Preservation, Department of the State. Each of them shared their knowledge and expertise on cultures and populations in the Gulf of Mexico Region and provided valuable contact information for other participants. The MMS thanks the following for their written and editorial contributions to this workshop report: Diane Austin, Kimberley Cook, Asha Luthra, Harry Luton, Diana Olien, Kristen Strellec, and Barbara Wallace. The University of New Orleans, Office of Conference Services, was the contractor responsible for the meeting coordination. We would especially acknowledge the patience and assistance of Patricia Arteaga and the students that assisted in each session. The staff of the Monteleone Hotel were accommodating to our requests. Support funding is provided through the MMS Environmental Studies Program.

CHAPTER 1

WORKSHOP ON SOCIOECONOMIC RESEARCH ISSUES
FOR THE GULF OF MEXICO OCS REGION

A workshop on social and economic topics related to the oil and gas industry was hosted by the Minerals Management Service (MMS), Gulf of Mexico OCS Region (GOMR), February 3–5, 2004, at the Monteleone Hotel in New Orleans, Louisiana.

Workshop Objective

The objective was to provide guidance for studies planning for the next five years. The workshop began with the premise that much has been accomplished during the last decade of MMS socioeconomic research and that it is time to assess future directions for the program. The workshop sought advice, rather than consensus, on key questions, useful analytical approaches, and critical information needs. While the workshop was intended to identify specific information needs, MMS was seeking "strategic thinking" about issue selection, methodologies, and assessment goals. Issues related to specific socioeconomic modeling techniques were not addressed by this workshop.

Workshop Structure and Schedule

The two-and-a-half-day workshop was divided into three parts (see Appendix B, Agenda). The first half-day plenary session included all workshop attendees and consisted of introductions, presented papers and discussions designed to acquaint participants with the goals and procedures of the workshop. The paper, "Social Impact Assessment and Offshore Oil and Gas in the Gulf of Mexico" by Harry Luton and Rodney Cluck shows that standard approaches to social impact assessment that focus on individual projects and demographic effects are inappropriate for the Gulf's oil industry. Richard Hildreth's paper, "Legal Requirements for Social Impact Assessments under NEPA," reviews court decisions regarding the adequacy of impact assessments and concludes that a great deal of flexibility in approach and content exists, particularly in the assessment of the "no action alternative." Tyler Priest's "The History of the Offshore Industry in the Gulf of Mexico" describes evolutionary changes within the offshore industry in the context of their onshore social and economic effects (see Appendix D, Presented Materials). The plenary session concluded with an organizational meeting for the breakout groups.

The second half of the first day and the following day consisted of three concurrent breakout groups, of approximately 20 members each, organized around broad assessment categories:

- Industry Trends and Dynamics in the Gulf of Mexico Region, chaired by Kristen Strellec and co-chaired by Diana Olien, Ph.D.

- Community-level Impacts of Oil and Gas in the Gulf of Mexico Region, chaired by Asha Luthra and co-chaired by Barbara Wallace

- Cultural Impacts of Oil and Gas Activity in the Gulf of Mexico Region, chaired by H. B. Kimberley Cook, Ph.D. and co-chaired by Diane Austin, Ph.D.

Each breakout group conducted brainstorming sessions focused on issues or questions related to the group's topic. There were three separate brainstorming sessions (Tuesday afternoon, Wednesday morning, and Wednesday afternoon). Each brainstorming session followed a similar format: (1) the session chair presented the group with an overall problem statement and/or set of general questions; (2) in order, each participant either contributed an idea or "passed"; (3) each idea was recorded on a flip chart; (4) as necessary, the session chair refined or reframed the general questions; (5) after the group had exhausted ideas on a problem statement or question, it categorized similar comments into emerging themes; (6) the group then prioritized the categorized comments or themes and formulated research topics; and (7) individual participants completed one or more of the Proposed Study forms provided by MMS. While this was the general brainstorming framework, each group varied in terms of how many steps of the brainstorming process were completed by the end of the session. Some participants submitted Proposed Study forms in the week following the workshop; some study ideas were submitted by more than one participant (see Appendix C). As part of the brainstorming process, each breakout group session was tape recorded. Formal written notes were taken in some, but not all, of the breakout groups.

The last half-day plenary session included two papers. Robert Winthrop described social impacts associated with oil development managed by the BLM and approaches the agency is taking to assessing and mitigating them. Donald Callaway discussed U.S. Park Service research on land use patterns among rural Alaskan communities to demonstrate techniques for formally identifying and measuring sociocultural impacts. The plenary session concluded with presentations from the three breakout groups and a synthesis discussion.

The intent of the workshop organization and its emphasis on brainstorming was to encourage strategic thinking about issue selection, methodologies, and assessment goals for socioeconomic research sponsored by the Gulf of Mexico Region. Chapter 2 discusses the key crosscutting issues identified by the participants. Chapters 3 through 5 provide the discussion papers presented during the first plenary session. Breakout group discussions and findings are presented in Chapters 6 through 8.

Workshop Participants

Workshop participants are listed in Appendix A. The workshop included MMS staff, invited participants, and other public participants. All attendees participated in the breakout groups. To facilitate the work of these groups, MMS sought to ensure that each had a pool of participants that, as a group: 1) exhibited disciplinary diversity; 2) exhibited a diversity of research interests including some with no previous experience with the GOMR OCS; 3) contained significant expertise within the research traditions addressed by group; 4) contained sufficient knowledge of socioeconomic impact topics addressed by the group; and, 5) contained sufficient knowledge of offshore oil, its operations and impacts. To that end, a large proportion of workshop participants had been invited to participate in a specific breakout group.

Background to MMS Studies Program

In the 1970s and 1980s, the GOMR conducted few social and economic studies. The Gulf's present-day approach to social and economic research dates from 1992. That year the National Research Council (NRC) reviewed the MMS studies program and the GOMR held its socioeconomic agenda setting workshop. The NRC argued that the greatest impacts from offshore petroleum activities have occurred in the GOMR over an extensive period from a wide range of OCS-related activities. For this reason, the NRC claims that the Gulf is a "natural laboratory" for the study of the industry's effects. Building on advice from the NRC, its Science Advisory Committee, several workshops and its many critics, the GOMR developed a broad, eclectic, comprehensive approach to the social and economic consequences of the program that includes studies of the industry and its dynamics, baseline descriptions, studies designed to better calibrate its economic and demographic projects, and studies of the industry's effects on communities, on recreation and other economic activities, and on human activities such as family life, crime and environmental justice. Since 1992, the GOMR has published the results of over 35 social and economic studies; almost 20 ongoing ones are yet to be published. Summaries and electronic versions of published studies are available on the web at:
   http://www.gomr.mms.gov/homepg/regulate/environ/techsumm/rec_pubs.html.
Descriptions of the ongoing studies are found at:
   http://www.gomr.mms.gov/homepg/regulate/environ/ongoing_studies/gom-se.html.

The workshop is also designed to address the many complex problems the GOMR has encountered as it has applied the NRC "natural laboratory" approach to NEPA to the Gulf offshore oil industry. Below are listed several of the "challenges" faced by socioeconomic assessments of the GOMR OCS leasing program:

- *The challenge of the baseline.* Under NEPA, an area unaffected by the proposal is the "baseline." In the Gulf, the industry has operated for decades. Since there is no environment "unaffected" by oil, in a sense, there is no baseline. In this situation, how does one separate the effects of oil from other regional, national, and worldwide trends?

3

- *The challenge of cumulative effects.* Since the industry is already in place, a lease sale's primary socioeconomic effect is to maintain the status quo. The State of Louisiana regularly complains that, since most socioeconomic effects of the industry are cumulative, sale-level assessments do not adequately assess the OCS program's real effects. What should constitute cumulative effects and how should they be separated from other regional, national, and worldwide influences and trends?

- *The challenge of the "affected area."* In addition to the complexity of the industry, the GOMR covers Texas, Louisiana, Mississippi, Alabama, and parts of Florida. Its 56 coastal zone counties and parishes include the extremes of social, economic, cultural, and institutional variation. How does one provide detailed assessments of industry effects within this vast expanse?

- *The challenge of the industry.* The industry is not well described and is actually a multitude of varied industries involved in finding, extracting, refining, and marketing petroleum-based products. Each industry has its own structure, economic dynamics, technologies, infrastructure requirements, labor organization and demands, and localized effects. How does one address this complexity and variability on regional and local levels?

- *The challenge of "local effects."* Many social and infrastructural effects are determined by specific local conditions—the unused capacity of a certain school district, the demands on a water system, or traffic on a specific road connecting a port and highway. How can the multi-county, sale-level assessments address local-level effects?

CHAPTER 2

SESSION ON INDUSTRY TRENDS AND DYNAMICS

This section of the Minerals Management Service workshop focused participant discussion on four main objectives:

1. Gaining a better understanding of what constitutes or defines the petroleum industry in the Gulf of Mexico Region (GOMR) and identifying key issues and trends in the industry as a whole as well as its specific sectors;

2. Gaining a better understanding of the key issues and trends associated with industries that support or are impacted by the industry in the GOMR;

3. Gaining a better understanding of the overall operating environment that the industry faces in the GOMR; and

4. Improving data collection and use.

From discussion over three workshop sessions, the following general themes or areas for study emerged: industry globalization and world market issues; examination of offshore and onshore industry structure, sectors, and players, individually and in terms of interaction; understanding projects and investment decision-making; issues relating to labor, both white- and blue-collar; and investigation of industries supporting or impacted by petroleum industry activity; improving data collection and use related to all these themes. What follows is an overview of how these general themes or areas for study were taken up in participants' discussion.

The Gulf of Mexico Region in a Global Petroleum Industry

Workshop participants identified a major area of study as the place of GOM offshore activity in a global petroleum industry, looking at the relative advantages and disadvantages of operations in the GOMR. In this context, participants wanted to know more about the resource base of the GOM and its reserves relative to other offshore basins. They raised questions about how profitable the GOM is compared to other offshore regions; how efficient GOM exploration is; how well the region competes at present for investor dollars; and what future prospects are in terms of the GOM's competitive position in these regards. The role of foreign players and state oil companies needs consideration as part of an assessment of the relative attractiveness of the GOM as an area for petroleum industry activity.

Participants wanted better definition of the place of the GOM in a world energy market. They noted differences, for example, between markets for oil and for natural gas and differences of pricing mechanisms in these areas. OPEC's influence on world petroleum prices needs examination, as does comparison of fiscal incentives and regulatory issues in the GOM and other petroleum producing regions.

Participants singled out a number of areas for study of impacts on GOM operations, which included the availability of cheap crude oil, the growing global market for liquid natural gas, changes in exchange rates, local/global practices, and alternative energy sources such as coal bed methane.

As participants noted, support industries are also part of the global market environment, and the GOM support sectors—fabrication, for example—must meet global competition. Here the issue is how well support sectors will be able to stay "ahead of the curve" and compete internationally; what would enhance their competitiveness internationally?

Characterizing Offshore Industry Structure and Players

Here the first focus was change over time in terms of those participating in offshore and onshore GOM oil and gas activity, in particular the increasing presence of independent firms—on the Shelf, in Deepwater, and in the global arena. In general, participants singled out the need to identify the various players: majors and independents, national and multinational, large and small in both offshore and onshore industry-related sectors. It was suggested that the Baxter (1993) study approach be replicated and updated.

Considering the major companies and independents, participants asked what the increased role of independents means in terms of expertise and level of specialization in operations, as well as in terms of spending for industry research and development. They raised the issue of interfirm relationships and how these have changed over time. They suggested that the increased role of independent firms may mean emphasis on cash return and enhanced shareholder value at the expense of improving industry infrastructure, asking whether independents rely on aging infrastructure without investing in updating or improving it.

Looking at independent firms, there is a need to characterize the companies involved in each sector of development, examining types of companies in each sector, size of companies, and degrees of specialization. There is also a need to look at the impact of mergers and acquisitions and at the degree of competition and consolidation among these firms. Barriers to entry, if any, should be identified, as well as factors that drive entry, exit, or consolidation in each sector. Different impacts of business cycles and differing vulnerability to price volatility among players and in sectors need examination.

Turning to specific upstream sectors, there should be study of geophysical service providers and problems they have faced with more data than there are buyers for, as well as the difficulty of selling buyers on new but more expensive technology. Consolidation in this sector was noted, as in drilling. With respect to drilling, it was suggested that the risks drilling contractors have assumed and their use of contract labor is worth study. More generally, participants were interested in service sector participation as joint venture partners, particularly with respect to risk sharing. It would also be useful to look at the changing views on the part of companies with respect to overall corporate mission and their identification with energy services as opposed to petroleum: do they see themselves as oil companies or energy providers? Here the example of BP as a self-identified "energy provider" going beyond petroleum was mentioned.

Turning to downstream, participants noted the importance of looking at downstream operations like refining and petrochemicals as they are affected by offshore activity. There is a need to know where regional production goes in terms of downstream processing and distribution; impacts on the petrochemicals industry of higher natural gas prices and what general trends seem to be emerging in that sector; consolidation in the downstream arena; changes in downstream demand for products; and environmental regulations as they relate to demand for natural gas. The future role of liquefied natural gas in meeting natural gas demand also needs study.

Projects

Workshop participants said there was a need for study of types of projects characteristic of current operations in terms of the Deepwater projects underway; changes over time in targets for investment; project objectives in terms of field size or additions to reserves; length of projects; application of newer technologies; and access to infrastructure. Better understanding of Deepwater projects is essential, especially with regard to their peculiarities and relation to price volatility. In general, however, there is a need to know how successful companies in the GOM have been at adding to/replacing reserves. Attention should be directed to differences between Shelf and Deepwater projects, particularly with reference to levels of capital investment, duration of projects, field development strategies, and demand for labor. The growing demand for natural gas and its impact on project planning and targets deserves attention; for example, notwithstanding the shift in activity off the Shelf to Deepwater, there is the possibility of finding deep gas reserves on the Shelf, a possibility largely unexplored. The search for giant fields is also a focus for investment. More data on the expenditure of companies over the life cycle of projects would be useful.

It is important to look at the infrastructure that supports projects, looking at what is in place, what it will take to maintain existing infrastructure, and what expansion or updating of infrastructure will be necessary. Participants observed that increased focus on natural gas will mandate changes and improvement in this area.

Participants expressed interest in the process of project decision making, in terms of how projects are developed, ranked in terms of rate of return, and budgeted, looking at changes over time in the relative importance of gas as opposed to oil. There is a need to look at technology issues in relation to projects; one example of this would be to look at the way new approaches to seismic data may rejuvenate projects in old fields. Technological questions raised the general issue of research and development. Participants asked where research and development was going on and how it was directed at reducing risks. They also identified questions of how to measure or assess the impacts of technological change and how technology is transferred across the industry.

It would be useful to know more about project finance and risk management in projects; how are risks assessed, and who bears them?

Labor

Workshop participants repeatedly raised labor-related questions in all sessions of discussion.

They identified a need to look at the characteristics of the GOMR industry labor force with respect to age, gender, education, place of residence, and living conditions, noting that the most recent survey directed at these issues was done in 1986 (Centaur Associates Inc. 1986). These characteristics needed to be studied in each sector of the industry, looking for differences, if any, between sectors. Differences in employment levels between sectors and the impact of technological change on each sector required study. Workforce quality issues were also raised, especially with reference to skills, productivity, and safety.

Particular labor issues identified included the "graying" of the workforce; the number of workers permanently leaving the workforce as a result of industry downturns; and the educational, training, staffing, and specialization levels of the workforce. These issues related to the need to look at changes over time in the GOMR oil and gas workforce, in general and sector by sector.

Many participants noted a need to examine the growing industry practice of contracting out a wide variety of operations in terms of what this has meant for workers. However, they also noted the difficulty of assembling data on the contract labor force, as well as in assessing its productivity. They raised the question of how the contracting out system has impacted and will impact jobs and the economy of the GOMR. In particular, they also asked whether there are differences between independent and major companies, and between sectors, with respect to contracting out. Contracting out in drilling was especially mentioned.

The growing shortage of oil and gas industry professionals, particularly petroleum engineers, received extended attention. Participants observed that there is a decreasing number of U.S. nationals, as opposed to foreign nationals, in the professions; they also mentioned the aging of geoprofessionals as a group in the petroleum industry. Recruitment problems like resistance to geographical mobility and attitudes toward the petroleum industry held by potential professional recruits exist in both upstream and downstream industry sectors. Competition in recruiting with the "dot.com" industries is an issue here as well. The quality and availability of university training programs is a concern in the supply of qualified professionals for the petroleum industry.

In terms of trained and skilled labor, the offshore industry was seen in competition with other industries like construction, refining, and petrochemicals. But the place of the offshore GOMR labor force in a global industry context is also worth study, particularly in terms of such issues as productivity, employment levels, and labor organization. With reference to labor organization, comparison to the North Sea and West Africa is desirable. Study of labor in all sectors and in a global context would be useful.

Overall, the most important issues relating to labor were the changing characteristics of the oil industry workforce over time; meeting industry needs for trained workers and professionals at all levels and in all sectors; contracting out; and worker productivity in a global industry.

Supporting and Impacted Industries

Turning to the industries that support oil and gas activity in the GOMR, workshop participants wanted to identify all the support sectors and ask how reliant these industries are on oil and gas as opposed to what they might do for other industries. Looking at the support sectors, there should be a thorough descriptive analysis of those involved, including size of companies, ownership of them, locations, consolidation and competition among them, labor mix, interfirm relationships (particularly between large and small firms), and how these characteristics have changed over time. In the context of this description, the problem of establishing the parameters of oil and gas industry support sectors was discussed. For example, if one goes into enough detail to ask where parts used by support industries come from, it is difficult to get relevant and accurate information.

The impact of industry activity on support sectors should also be studied by looking at how the size and duration of projects affect support sector mix, how support sector industries finance projects, and how oil and gas industry business cycles impact support industries/sectors. The place of support industries in a global petroleum industry needs examination, particularly with respect to whether support industries are competitive internationally. It was noted that foreign governments often subsidize their own players in support sectors. Sector by sector, what would enhance international competitiveness of the GOM support industries?

Attention should also be directed toward support sectors' use of contract labor, training of workers, and relation to local economies. The environmental impacts of support industry operations and the ways in which support industries are meeting the challenges of environmental regulation deserves study. Once again, in this area, attention should be directed to changes over time. In the environmental context, particular concern is the issue of coastal erosion and how leasing activity has impacted this over time. State policies and regulations' impact on support sectors should be examined.

In the context of support and impacted industries/sectors, workshop participants singled out several for specific attention. These included catering services, with study including local economic impact and labor mix; coring, logging, and geophysical services; the insurance and legal sectors, including changes over time in costs of insurance and the changes in the insurance sector in general; banking and financial institutions; environmental consulting and remediation; and housing and real estate. Ports were identified as an area needing examination, including both public and private facilities, labor in ports, and infrastructure necessary for port functions. However, the difficulty of obtaining relevant data on port activity was mentioned as a formidable problem.

When workshop participants looked at the interaction of industries in the diversified economy of the GOMR, they identified labor as an area in which industries involved in oil and gas activity compete with others outside this arena, specifically for skilled, trained workers. Here it would be useful to know the degree of specialization required of workers and the extent to which they can and do retrain for work outside the petroleum industry. Within the industry, one should look at competition for labor between offshore and onshore sectors, paying particular attention to construction, refining, and petrochemicals.

Turning to competition between petroleum-related industries and those not connected with oil and gas, participants saw a need to contrast oil and gas activity to activity in such areas as tourism and commercial fishing in terms of overall economic value and environmental impacts. What is the magnitude or intensity of use of the GOM on the part of non-petroleum industry users? Here recreational use, as, for example, beaches and rig fishing/diving would be included in investigation. How does the petroleum industry compete for space in GOM waters with the military, as well as with fishermen?

Impacts of petroleum industry activity on non-petroleum industries/sectors are worth examination, with specific attention directed toward real estate and the housing market, tourism, and financial institutions. Study of impacts on the petroleum industry of federal and state planning already underway is needed; MMS needs to know what other governmental agencies are planning.

Land loss was identified as a critical problem in the GOMR. Participants raised the question of the impact over time of lease sales and oil and gas activity on coastal erosion. Specifically, there is a need for more information on how the current rate of land loss impacts OCS activities, looking at industries/sectors both connected and separate from oil and gas—ports, coastal infrastructure, commercial/recreational fishing and diving, for example.

A final impact area related to intraregional shifts over time in the locus of business activity. More especially, what will be the impacts of an increasing concentration of petroleum industry planning and operations management in Houston as opposed to New Orleans or Lafayette? What will the focus on Houston mean for the GOMR as a whole? To what extent will other GOMR cities or communities continue to compete successfully with Houston as bases of petroleum-related operations?

Data Collection

To approach the issue of data collection, workshop participants began by identifying what data, ideally, would be desirable to have. These included exploration and development budgets for individual companies; data on fixed and variable costs; lease by lease actual expenditures; costs by water and well depth; employment records; data on contract labor; amount of refinery use of imported oil; state and local finance data; data on use of imported liquefied natural gas; data on public and private port facilities; and data on risk sharing. Participants noted that data collected in Centaur (1986), individual labor force information by zip code, would be useful to update.

The practical barriers to acquiring much of this data, however, appear to be challenging. Here workshop participants heard from Zeta Rosenberg who has been trying to collect data on a wide range of GOMR OCS industry spending and costs over the last five years.

In organizing her data collection project, Rosenberg took a number of steps to try to ensure success. She made use of regional economists and statisticians, and hired people who knew people in the industry. To be inclusive, the survey went well beyond exploration and production to include support companies, including seismic services, transportation providers, and fabricators. To try to improve response rates in what was a voluntary survey, surveyors sent out letters to everyone and followed up with weekly phone calls. Notwithstanding these follow-up efforts, none of the contacts responded within the first month of survey release. In the end, the response rate was low, averaging about 20% of those contacted. In fact, in some areas, response rate was so low that information use would have violated confidentiality. In others, responses were erroneous or simply lacking; no one, for example, supplied data on labor turnover rates. Information about contractors was especially hard to get.

So what should be done about future surveys? Rosenberg suggested that in the future, MMS data gathering should coincide with industrial census data gathering. There probably need to be fewer and simpler surveys; as another workshop participant said, "Simplify, simplify, simplify." He added that when he had conducted a survey, he had gone out to companies and done the work of data collection himself rather than expect companies to perform it. Rosenberg and others noted that it must be clear to industry respondents how responding benefits them—what they will gain from participation. Other workshop participants noted that the amount of time required to assemble and complete responses to the survey, three days, was excessive, and that it would be unreasonable to expect smaller firms to give that much time to a voluntary effort. Industry volatility—firms merging or going out of business—also made it difficult to get information. In passing, it was noted that industry organizations like the American Petroleum Institute seem to fare little better at data collection.

Overall, workshop participants said it would be desirable to decide which data really was necessary and important to collect, as well as to explore better methods of getting data. In particular, industry participants should be able to see how cooperation would be to their benefit. It would also be desirable to make better use of information collected by state and local agencies, though this raised the problem of separating out the impacts of state and federal policies and regulation of offshore. More analysis of data MMS already collects, as well as a need for more user-friendly data, was also suggested.

References

Baxter, V.K. 1993. The Effects of Oil Industry Restructuring in Louisiana. *In*: S.B. Laska, V.K. Baxter, R. Seydlitz, R.E. Thayer, S. Brabant, and C.J. Forsyth, eds. Impact of Offshore Petroleum and Production on the Social Institutions of Coastal Louisiana. Prepared by the University of New Orleans, Environmental Social Science Research Institute. New Orleans, LA: U.S. Dept. of the Interior, Minerals Management Service, Gulf of Mexico OCS Region. Cooperative Agreement 14-35-0001-30470. OCS Study MMS 93-0007, pp. 43–76.

Centaur Associates, Inc. 1986. Indicators of the Direct Economic Impacts due to Oil and Gas Development in the Gulf of Mexico: Executive Summary; Volume I: Narrative; Volume II: Exhibits and Data. 3 vols. New Orleans, LA: U.S. Dept. of Interior, Minerals Management Service, Gulf of Mexico OCS Region. Contract 14-12-0001-3-178. OCS Study MMS 86-0014, 86-0015, and 86-0016. 33, 100, and 706 pp.

CHAPTER 3

COMMUNITY-LEVEL IMPACTS OF OIL AND GAS ACTIVITY
IN THE GULF OF MEXICO REGION

This section of the Minerals Management Service workshop focused participant discussion on three main objectives:

1. Identifying types of fiscal and infrastructure concerns facing oil-involved communities in the Gulf of Mexico Region (GOMR).

2. Defining the current socioeconomic effects of the oil and gas industry on communities in the GOMR.

3. Defining the current socioeconomic effects of the oil and gas industry on urban areas in the GOMR.

Each of the three workshop sessions addressed one of the above objectives. Six general themes emerged after the brainstorming ideas and discussion from each session had been synthesized. The themes included: 1) understanding community dynamics; 2) understanding the community public administration structure; 3) understanding adaptation and accommodation strategies; 4) understanding the significance of environmental impact statements; 5) using appropriate forms of research methodology; and 6) examining industry-specific issues that affect communities in the region. We provide an overview of how these themes emerged in the workshop sessions below.

Community Dynamics

Stakeholder Groups

Workshop participants identified the need for MMS to better understand the role of different groups in a community. For example, how do we identify the stakeholder groups in each community? How do these stakeholders directly or indirectly affect the community's relationship to the oil and gas industry? The participants discussed how the power structure of various communities needed to be understood to address marginalized groups negatively affected both economically and socially. Participants also believed that identifying the voice of a community was necessary to understand the nature of the relationship between the community and the offshore industry.

Social Networks and Social Capital

Participants wanted to gain more insight into community social networks. In other words, how do the formal and informal interpersonal relationships between community residents affect the community/industry relationship? Understanding community networks can help MMS better understand why the industry has different socioeconomic effects in different communities across the GOMR. Some participants also wanted to know whether social networks coincide with geographical and/or political units. Answering this question would help us better define communities that are affected by the oil industry. Some workshop participants also wanted to compare the amount and type of social support or social capital in oil-involved communities to communities that are not involved in offshore oil. Social capital is the structure of interpersonal relations and institutional linkages that create social ties that bind a person to the community and that enhance an individual's life chances (Coleman 1988:98; Bourdieu 1986). Social support has implications for health, education, family, employment, and other life outcomes. It is important to examine network structures in a community to understand and predict how people adjust to difficult and changing situations such as cycles in the oil and gas industry. Studying social networks also allows us to measure the amount of solidarity among residents, which can be an indicator of community satisfaction.

History of Communities

Finally, participants believed that historical analyses of communities needed to be incorporated in future studies. It is one of the only strategies to understanding of the cumulative effects of the oil industry on different communities across the GOMR. Historical analyses also help us better understand industry effects that are long-term and complex. The importance of an historical perspective came up in each of the three community impact sessions.

Community Public Administration Structure

Fiscal Structure

Workshop participants wanted to know more about the disbursement of tax revenues in communities. For example, what are the direct and indirect mechanisms that affect the fiscal structure of communities? Where does the money come from? Who controls its disbursement? Is the money that results from offshore oil development dispersed equally across groups and communities? Are there differential tax implications and demands for various industry sectors? Answers to these and similar questions would most likely explain the economic dependency of communities on different sources of revenues. In turn, this information would improve our ability to predict potential community effects of the industry.

Although examining the fiscal policies of specific communities was discussed, it is also necessary to compare the policies across states in the GOMR. One possible way to research these differences is to systemically analyze the annual state budgets for the five states in the GOMR.

Participants wanted to compare the delegation of state revenues to local communities across the region. Others believed that it was particularly important to differentiate transfer funds from direct funds within communities to better understand existing fiscal structures.

Strategies to Pay for Infrastructure

Fiscal structure and soundness are directly linked to community infrastructure. The various strategies to pay for community infrastructure were discussed. Participants believed that the amount of tax base diversification in a community was important in gaining pertinent information about potential impacts. In addition, fundamental industry changes should be more closely examined to fully understand a community's ability to pay for infrastructural needs. Others suggested that we look at different mechanisms that communities may be experimenting with to deliver public services (e.g., privatization).

Carrying Capacity of Infrastructure

The next topic focused more directly on the carrying capacity of various infrastructural resources. For example, what is the preexisting capacity of a community to respond to oil industry changes? Much of the discussion addressed the carrying capacity of physical forms of infrastructure such as ports and roads due to increased oil-related activities and the in-migration of workers. However, some discussants believed that it was important to look at public services in a broader way. For example, churches can also play an important role in a community's reaction to industrial change (e.g., collapsing oil or gas prices, the closing of a plant, etc.). This broadened concept of infrastructure allows for more flexibility and a higher probability of addressing impacts.

Adaptation and Accommodation Strategies

Community

Differential community response to oil and gas industry activity came up repeatedly across all of the workshop sessions. The participants believed that it was important to understand the level of dependency of various GOMR communities on the industry. For example, to what extent do communities redefine themselves to fit the industry? In turn, how do industries shape communities to fit their needs? Do you build an oil town or does the oil company build a town to fit it? This process would affect the types of public services and infrastructure that would be available to residents. Some participants argued that even if different communities reached similar relationships with the oil and gas industry, it is more important to understand the process of how they reached that relationship than the type of relationship itself.

Some of the study questions included: If one or two platforms are decommissioned, what effect does that have on the communities that have been supplying them? What is their nature of industry dependency? In addition, how does industry job outsourcing affect different communities and states in the region?

In the last workshop session, the group primarily focused on the adaptation and accommodation strategies of major urban centers in the GOMR. Several study ideas were proposed to conduct business case studies that would examine the impact on a city such as New Orleans when an oil company moves to another city such as Houston. Although some believed that the loss of approximately 1,000 white-collar jobs would be minimal to a large city's economic stability, they predicted that significant impacts were more likely to be found in civic pride and the overall reputation of the city. Following a social constructionist perspective, it could be more of a matter of perception than real change. One way to examine such industry impacts is to trace displaced employees from one location or position to another. For example, did displaced employees move to another city? Did they stay in the same location and change occupations or did they choose to start their own businesses? Another possible impact could be seen in real estate markets. If displaced workers were clustered in particular neighborhoods and moved at the same time, it could have significant impacts on property values and schools. Schools may lose some of their most active and involved parents. Community organizations and churches could also lose some of their most active members. Although these effects would not be as noticeable as in a small community, there is little existing information on industry impacts in large cities.

Many of the participants believed that a few case studies of oil and gas companies should be conducted to determine if the findings were viable enough to pursue further. The importance of conducting historical case studies of oil-involved communities and major oil and gas companies was mentioned consistently. Participants also believed that the level of significant change needs to be addressed. For example, a series of case studies of oil companies might find a number of important changes. However, an economic model could determine that there isn't any statistically significant impact at the macro-level. It is important to examine effects at different levels of analysis (business, household, city, county, neighborhood, etc.) because the lack of significant impacts at one level does not necessarily indicate that the effects won't be significant at another level or cumulative over time.

One way to better understand a community's adaptation and accommodation strategies is to examine the local social economy (Tolbert et al. 1998; Tolbert et al. 2002). To what extent is there a local business constellation? Is the cluster primarily local or is it dominated by outside or corporate interests? It is also important to determine the overall productivity and efficiency of the constellation as well as understand the mechanisms that make it work. What are the attributes of communities that support the development and actions of local business constellations? It is necessary to examine local social economies of various communities because it can be a strategy of adapting to shocks in the environment. Consequently, these local businesses can "become buffers for communities that insulate them from global forces" (Tolbert et al. 1998:407). It is possible that port cities are structured in a way that helps them adapt to all types of economic development or decline—even those not directly tied to the oil and gas industry. Port cities may be resilient in a variety of ways due to the actions of local businesses.

Individuals and Households (Work and Lifestyle)

Considering adaptation and accommodation strategies, participants also wanted to focus on individuals and households. Some participants wanted to know more about the economic household strategies of those working in the oil industry. For example, community effects extend beyond job creation or job loss. Effects can be far more subtle and indirect in that people can lose opportunities to maintain a certain lifestyle. Some people need formal economy options to maintain other lifestyle patterns (fishing, hunting, farming, etc.). Determining the economic tipping point is important in better understanding people's decisions. For example, it would be useful to know how much income someone working in the oil and gas industry needs to maintain their alternative work activities such as shrimping and fishing. Would these lifestyles disappear entirely without the availability of offshore oil and gas employment? What are the benefits and burdens of industry involvement on families in the region?

The Environmental Impact Statement

Workshop participants discussed the utility of Environmental Impact Statements (EIS)—who uses them and how. In addition, some participants wanted EISs to address the implications of specific industry-related effects. Others thought these effects needed to be defined better. For example, the relativity of effects (positive and negative) needed to be addressed. Thresholds of significant effects should also be determined.

Research Methodology

One of the brainstorming ideas that received the most support was the promotion of monitoring. The EIS was suggested as a tool to systematically evaluate community changes. One of the most important tasks to accomplish effective monitoring is to develop a small set of baseline universal indicators. These indicators could be used consistently over time and place to study change. Measurement of these common social indicators in surveys, combined with case studies of the unique stories that shape different communities, would give MMS a much better sense of community effects across the entire GOMR.

More specifically, case studies would be performed to identify the set of sensitive indicators that will be used. These same variables should also appear in larger scale aggregations using county-level census data and other secondary data sources. After these steps are completed, long-term community monitoring can be used to track changes at multiple levels of analysis and describe cumulative impacts.

In addition to monitoring, workshop participants emphasized the utility of various types of data (e.g., primary, secondary, cross-sectional, and longitudinal) and research methodologies (e.g., ethnographies, case studies, historical case studies, surveys, network analysis, and GIS).

Industry-Specific Issues

Industry Impacts on the Community

Several study ideas focused on the impact of industry changes on communities. One of the infrastructural changes occurring in many ports is the deepening of navigation channels. Deeper navigation channels are justified by creating and maintaining jobs as a result of construction projects or fabrication facilities. Is channel deepening occurring in specific communities? It would also be useful to examine the potential socioeconomic effects resulting from the competition between ports and shipyards that support OCS activities and offshore module fabrication. For example, how does channel deepening affect local and state tax revenues and overall economic development?

Some workshop participants also wanted to know how oil and gas activity becomes concentrated in certain areas like Port Fouchon. What are the consequences of this concentration on communities (e.g., roads, highway access, zoning, new pipelines, etc.)? Given that new LNG plants will be constructed along deepwater ports, how will this affect the current pipelines? What is the pipeline capacity through Louisiana coastal wetlands? Finally, some participants wanted to study the impact of decommissioning and onshore waste disposal on communities.

References

Bourdieu, P. 1986. The Forms of Capital. *In* Handbook of Theory and Research for the Sociology of Education. J.G. Richardson, ed. Greenwood Press, pp. 26–65

Coleman, J.S. 1988. Social Capital and the Creation of Human Capital. American Journal of Sociology 94: S95–S120.

Tolbert, C.M., T.A. Lyson, and M.D. Irwin. 1998. Local Capitalism, Civic Engagement, and Civic Welfare. Social Forces 77:401–27.

Tolbert, C.M., M.D. Irwin, T.A. Lyson, and A.R. Nucci. 2002. Civic Community in Small-Town America: How Civic Welfare is Influenced by Local Capitalism and Civic Engagement. Rural Sociology 67(1): 90–113.

CHAPTER 4

CULTURAL IMPACTS OF OIL AND GAS ACTIVITY IN THE GULF OF MEXICO REGION

While social impact assessment often mentions effects on culture, little is generally done to address this issue. The cultural impact group session was organized with the intent of breaking new ground. The goals of this session were to:

1. Identify which populations and cultures in the region should be studied;

2. Describe and compare the cumulative affects of the offshore oil and gas industry on cultural groups in the region;

3. Uncover possible similarities in the ways that industry sectors have affected features of culture throughout the region;

4. Determine if universal features of culture can be used as predictive units of analysis in identifying industry impacts; and

5. Examine social impact assessment issues with a focus on the topic of environmental justice.

To develop a basis for discussion, during the first breakout session, one of our participants, Dr. Don Callaway, presented his research on social network analysis in Wales, Alaska. This study was meant to provide the group with an example of how impacts on a feature of culture could be studied. Participants were then asked to comment on the following: 1) Could this study be used to investigate cultural impacts in the GOMR? 2) If so, how would it need to be altered? 3) Could it be used to study sub-cultural groups? 4) What other types of research projects could be developed for the GOMR? During the workshop the greatest emphasis was placed on the development of general study themes and research directions. While our group was diverse in terms of area of expertise, our regional experts came from the states of Louisiana and Texas. Thus, while many of our study themes can be applied to the entire GOMR, our focus was on the Central and Western areas.

General Questions and Study Themes

The following general questions were addressed during the cultural impacts session:

- How do we identify cultures? Which ones should be considered for future MMS studies?

- What are industry variables that impact culture?

- How does the industry impact features of culture?

- How do groups respond and adapt to changing environments?

- What are the environmental justice issues we need to consider?

How do we identify cultures? And which ones should be considered for future MMS studies?

At the outset of the cultural impacts session participants brought up the problem of identifying cultural groups for analysis. Much of the discussion was on the need to understand changes in oil-affected communities such as Morgan City or New Iberia, often defined as counties or parishes. Participants noted that impacts on those communities involved with fabrication and support would differ from those due to downstream activities such as refining. Participants suggested that we need to go beyond the study of counties and parishes and look at specific cultural groups. The following in the state of Louisiana were suggested: the Isleños, the Chitimacha Indians of Saint Mary's Parish, and the Houma Indians of Terrebonne and Lafourche Parishes.

Group participants then suggested that regional cultures may be more significant in understanding industry impacts than are specific ethnic groups within the region. Some of the regions that participants suggested included: the Creole/Gulf Coast culture; South Louisiana, North Louisiana, and New Orleans. Many participants thought this approach was particularly relevant in "Cajun Country" since it had been defined, in part, by participation in the oil industry. In describing these regions it was suggested that we compare the cultures of today to those prior to 1937. Additional consideration should be given to generational issues across the region that are related to the oil industry. Thus, workshop participants identified various types of groups as cultures. In relation to the oil business, geographical differentiation from the coast outward and from key points such as concentrations of refineries, support or fabrication facilities may be the most relevant units of analysis to consider.

Cultural groups could also be described based on shared features of culture such as language, subsistence practices, and economic organization. One group that has been overlooked is the small scale economies of south Louisiana that continue traditional subsistence practices. These groups often use traditional subsistence practices and harvest areas. They have often been viewed as "economically-challenged" because these activities are not recorded by the U.S. Census. Participants suggested that it is important to understand the small-scale economies of the coast for two reasons. First, many of these populations still rely in part on their traditional subsistence economies such as crabbing or fishing as part of their survival strategy. And second, they make a contribution to the heritage, cultural traditions, and economy of the larger system. Session members thought that identifying traditional contributions to local ways of life, and identifying the impacts of the oil and gas industry on them are issues that should be addressed.

Our participants pointed out that these traditional coastal communities of Louisiana are internally structured and somewhat closed to outsiders. In describing these communities we need to identify groups of people within them to give us a more accurate description of their internal organization. An insider's point of view will help us more accurately describe industry impacts. We also need to look at their varying attitudes concerning the oil business. In addition, cultures often define themselves by what they are not. It was suggested that we see how communities define themselves along these lines. Group members suggested the following methods for obtaining this information: 1) the use of ethnographic field methods; 2) the collection of historical narratives at the community level; 3) and open and structured interviews. These methods would allow us to more accurately identify stakeholders in the region.

A problem we need to avoid in describing cultural groups is treating them as fixed, monolithic entities. Cultures are by definition adaptable. Labeling them as homogeneous, fixed units might lead to the possible problem of their not being considered as legitimate by the courts and other official bodies.

A suggested approach to the problem of cultural effects was to describe cultural niches shaped or influenced by the industry. For example, one could identify cultures based on shared themes and events such as the Petroleum Festival. One could trace the evolution of this festival to show the changing cultural influences in the area.

Another idea was to identify occupational communities. Participants suggested that we could understand the worker culture by describing their roles, positions, training needs, and stresses. This culture could be assessed for the present as well as for future industry trends. Participants also suggested that we examine multigenerational issues that transect these communities. In addition, we should investigate the common concerns that are related to occupational groups such as lifestyle issues.

Other suggestions included looking at how land developers and coastal land development are driving culture change in the Florida panhandle. There, local cultural values are being redefined by relationships between an industry and in-migrating groups. Similarly, participants asked how trends in various petroleum industry sectors will change the profile of migrant workers that come into the state of Louisiana and into the Gulf of Mexico in general. In addition to the affected area, we need to describe the groups who make long commutes to work in various industry sector jobs and how this job pattern affects their cultures, communities, and hometowns. It would also be beneficial to focus more attention on the workers and their work environment both offshore and in support industries.

Much anthropological research has focused on traditional, small-scale societies. In the past decade many oil corporations have moved out of southern Louisiana and have established their corporate operations in Houston, Texas. To provide a more balanced perspective of industry impacts it could be useful to obtain information on the culture and activities of white-collar and other workers in this geographic location.

In his plenary session paper entitled: Legal Framework for Minerals Management Service Socioeconomic Studies in the Gulf of Mexico, Dr. Richard Hildreth suggested that we examine environmental impacts in the GOMR that originate from neighboring countries outside of the boundaries of the United States. Expanding on this idea, our group participants suggested that we also consider cultural impacts in the Gulf that emanate from Latin American and Circum-Caribbean countries. To gain a more thorough understanding of migrant populations impacted by industry activities in the United States, it would be useful to view them from a more relativistic perspective. For example, while the partial integration and participation of migrants in the U.S. system might be seen as a negative impact, migrants themselves may prefer to maintain their primary socioeconomic base and support networks in their countries of origin.

During our final workshop discussion it was suggested that we describe corporate cultures within the industry. Impacts can differ based on the ways in which different companies are structured and operated. Understanding these differences might give us a better understanding of impacts on workers, their families, and communities.

At the end of our discussion of cultural entities it was suggested that we consider the financial practicalities of our proposed study ideas. As MMS's funding is limited, it is important to rank studies in terms of their relevance to NEPA and their utility as support for improving Environmental Impact Statements, mitigation, and monitoring.

What are industry variables that impact culture?

Participants questioned how future trends in the oil and gas industry will modify baseline cultural effects. Some of the trends discussed included: 1) The move to deep water as a response to royalty relief; 2) The increase in sophisticated technology and educational backgrounds of the worker population; 3) The shorter lifecycle of projects, but the longer duration of technical experts working offshore; 4) The globalization of the industry with operations in Qatar, Nigeria, Brazil, Trinidad, etc.; 5) The importation of equipment and supplies from international firms; 6) Royalty relief for shallow-water, deep gas operations; and, 7) Liquefied Natural Gas (LNG) operations in the Gulf.

In relation to Gulf-wide trends, participants asked the following questions: 1) How will new energy resources, equipment, and tools affect impacts? 2) How will the workforce, their needs, and training requirements be affected? 3) Where will new workers come from? 4) How will we identify stakeholders? 5) What types of industries will develop in this country to support the global activities of the oil and gas industry abroad?

In discussing these trends participants brought up the fact that the oil industry has often helped maintained cultures in the region. Therefore, it was suggested that we examine how the potential

22

absence of industry activities will impact cultures in the region. This discussion emphasized that research should include both negative and positive impacts as well as direct and indirect affects of industry activity in the Gulf.

One participant observed that we need to look at the difference in impacts between the states in the GOMR. While comparing Louisiana and Texas it was pointed out that Texas has been better able to absorb the benefits of oil and gas activities. The question was asked, how do state-level infrastructures vary in their ability to absorb and benefit from industry sector activities?

How does the industry impact features of culture?

In addition to identifying cultural entities we were also interested in describing how industry sectors impact universal features of culture, including subsistence technologies, economic systems, social organization, gender relations, informal networks such as associations and interest groups, and belief systems.

In looking at possible uniformities in the ways that features of culture are affected it was suggested that we reexamine the topic of offshore work schedules and their effects on communities in the region. This topic was recently addressed by an MMS study entitled "Social and Economic Impacts of Outer Continental Shelf Activities on Individuals and Families" (OCS Study, MMS 2002-022). This study was conducted in Morgan City and New Iberia, Louisiana. Participants suggested that we compare the 14-days-on and 14-days-off work schedules of offshore workers throughout the region to see if the impacts on family and community members are similar. Another idea was that we compare differences between the Gulf of Mexico and Alaska regions. By comparing particular universal features of culture we might be able to develop predictive models in assessing industry impacts.

One method for understanding impacts on features of culture would be to use social network analysis. This approach would work well for closed communities in the Gulf such as Vietnamese populations but would have to be modified if it were to be applied to larger and more diffuse cultural groups such as the Cajuns of Southern Louisiana. Along these lines it was also suggested that we use some form of social network analysis to identify an impacted phenomenon within the culture such as certain types of plants or indigenous food sources. After identifying them we could then describe how they have been impacted by oil and gas activities.

Group discussants developed a larger and more inclusive model that describes industry impacts on cultures by using historical eras and project lifecycle phases as the independent variables. The four historical phases would include industry activities: 1) on-shore, beginning in the early 1900s; 2) on the shelf in shallow water beginning around 1953; 3) at present in deepwater; and 4) and future activities in ultra-deepwater. Each era would then be described in terms of the project phase, including its activities and resource needs. Impacted features of culture such as social structure could then be compared across eras based on industry activities. As an example, one could compare differential activities related to production with a focus on labor requirements and work schedules. By comparing this information across eras one could describe the cumulative impacts of this phase of the industry on social structural systems in the communities where workers come from as well as the networks that develop in the occupational cultures of workers on the rigs.

Participants suggested additional features of culture where we should describe impacts. One idea was to look at the educational systems in the GOMR. How have formal and informal educational systems been impacted? How will educational requirements and training be altered to meet future industry demands?

The cultural impact group participants brought up new and interesting features of culture that are worthy of investigation. One of these was material culture. For example, in the state of Louisiana it would be interesting to describe how the infusion of cash and oilfield materials have been used to alter yardscapes and internal house features. Along these lines participants suggested that we describe cultural innovation in other areas as well. Specifically, how have groups in the GOMR successfully incorporated, altered and utilized impacts from the oil and gas industry? Again, this was part of the emphasis on positive contributions of the industry to cultural expression.

Additionally we should consider industry impacts in the areas of generational differences, language, gender, and social structure. On the topic of generational differences we need to compare the differential experiences and degree of commitment for different generations involved in the industry for the GOMR.

Other discussions about features of culture included the suggestion that we examine small-scale economic systems and the local environment. From a methodological point of view, participants suggested that we use more ethnography, historical narratives, and insider types of approaches.

How do groups respond and adapt to changing environments?

Coastal erosion, a major concern of the state of Louisiana, was brought up by the cultural impact session participants. Because erosion is a significant factor in local culture change, participants felt that we need to consider its impacts in any understanding of the oil industry's effects on coastal populations and cultures in Louisiana, and possibly elsewhere. Participants asked: What are population shifts in response to land and job losses? What is the impact of land loss on the sense of family, community and place? How will traditional small-scale subsistence economies be altered? When people are displaced from their homes and from industry jobs, what are the implications for family and social networks? In particular, how are extended family members affected? What are population movements like in terms of age, race, and ethnicity? How is language affected? How is participation in the culture affected by population movement due to land loss? How has traditional knowledge of specific environments been altered? When we look at land loss, what are the cumulative effects? What percentage is due to industry activities?

Some participants brought up the issue of climatic change. In particular they wanted to know which issues might be relevant. What types of impacts will there be on coastal communities?

Participants then suggested that we conduct an investigation of historical changes in community identification with the oil industry. The hypothesis might be that significant turning points in the industry correspond with notable and measurable changes in the way residents buy into and "own" the oil industry as an appropriate locus of regional identity and sense of place. For example, as the industry moves into deep water and labor demands change, communities respond in measurable ways with greater distance from the industry. As erosion issues deepen, perhaps identifiable sectors of the community realign their cultural identities and associations with the industry.

What are environmental justice issues we need to consider?

Based on the requirement that the MMS address environmental justice issues, participants recommended that we develop a mini-atlas of low income and minority populations in Louisiana and elsewhere in the GOMR. The three most prominent groups to be considered would be African Americans, Hispanics, and Native Americans. A tiered geographic approach could be used to expand beyond coastal communities. A difficulty with this approach is that it may ignore important sub-cultural group distinctions. Group members also reiterated that many of these issues relate to impacts on all inhabitants of the area, not just on low income and minority populations. For example, how have non-minority males been adversely affected by recent industry trends since they compose much of the industry's workforce?

Participants suggested two studies to expand on an on-going study by the MMS, entitled "Environmental Studies in Lafourche Parish, Louisiana." This study found that there is a possibility that minority groups (Houma Indians) could be disproportionately affected by OCS planning activities. Based on this finding, the first study would ask the following questions: 1) What are the consequences of possible disproportionate exposure and risk of exposure? 2) Can any specific negative impacts be documented, or even shown to be plausible? and 3) How do Houma traditional activities either diminish or enhance the disproportionate risk?

The second study proposed that we investigate the longer-term historical processes of settlement and population distributions along the Gulf Coast to consider 1) whether regions other than Lafourche Parish are also likely candidates for environmental justice investigations; 2) how the Houma came to be in the precise locations where they are now; and, 3) whether current resettlement issues that arise from coastal subsidence can be framed and understood in a more environmental justice focused narrative. Several participants emphasized that we should investigate this issues in terms of historical changes in community identification with, and participation in, the oil industry.

CHAPTER 5

CROSSCUTTING ISSUES AND QUESTIONS

This workshop was organized into three discussion groups that addressed different sets of questions. Each group consisted of individuals with widely differing backgrounds, interests, and experience with the offshore industry. Nonetheless, each group identified important questions and concerns that were also identified by the other groups. These crosscutting issues are summarized in this chapter. Instead of deciding at this moment which ideas might be funded and which put aside, the MMS plans to consider this entire spectrum of expert advice about issues, methods, and strategies as it develops its socioeconomic research agenda over the next five years. See Appendix C for the complete list of study ideas generated at the workshop sessions.

Industry

A recurring theme raised in all workshop sessions was the need for a better understanding of what constitutes, or defines, the oil and gas industry in the Gulf of Mexico Region (GOMR). A major issue for all sessions was the place of the GOM offshore industry in the global one. While, in the industry session, the emphasis was on the place of its activity in a global petroleum industry and energy market and the relative advantages and disadvantages of operations in the GOMR in terms of attracting investment dollars, all sessions wanted a better understanding of the scope and causes of the geographical shifts that have taken place within the industry sectors. For example, the concentration of corporate headquarters in Houston has social and cultural effects not only in Houston, but across the GOM region. Participants wanted a better understanding of all upstream and downstream industry-related sectors and emphasized change and the relative importance of change.

Turning to the industries that support oil and gas activity in the GOMR, workshop participants wanted to identify and thoroughly describe them on a sector by sector basis and ask how reliant these industries are on oil and gas as opposed to other industries. Specific support industries mentioned included catering services; diving; coring, logging, and geophysical services; the insurance and legal sectors; banking and financial institutions; and real estate. Ports were also identified as an area needing examination, including both public and private facilities.

Another industry characteristic that participants identified was the need to better understand the changes in corporate structure and individual project characteristics that have taken place over time and how these changes have impacted different stakeholders, including various cultural groups. Important structural issues included ownership (i.e. major vs. independents; foreign vs. domestic), firm size, and degree of specialization and vertical integration. Important project characteristics included 1) the move to deep water; 2) the increase in sophisticated technology and educational backgrounds of the worker population; 3) the increase in the complexity and capital investment required for projects; 4) the shorter lifecycle of projects, but the longer duration of technical experts working offshore; 5) the importation of equipment and supplies from international firms; and 6) the growing importance of natural gas and liquefied natural gas (LNG) operations in the Gulf. The cultural impact members would like to describe and compare effects on features of culture by using industry eras and project lifecycles as the independent variables to better assess cumulative effects of industry impacts throughout time and across the region.

Participants also noted the importance of studying corporate culture. It was suggested that we describe the different organizational cultures within the industry and how they impact the community. Issues included how individual firms view risk and risk management, investment in research and development for new technologies, labor relations and their overall role in the community. Some participants also wanted to know more about community dependency on industry. They wanted to know whether diversification in industry sectors affects a community's ability to adapt to change. Determining the nature of dependency on the oil and gas industry would allow for more precise predictive models of community effects.

It was suggested that we need to better understand the role of infrastructure. It is important to look at the infrastructure that supports projects, looking at what is in place, what it will take to maintain existing (and aging) infrastructure, and what expansion or updating of infrastructure will be necessary. Participants also emphasized the effects of industry change on the capacity of community infrastructures and the ability of communities to pay for and manage them.

Stakeholders

The second major theme that emerged across workshop groups was the need to identify and understand the various stakeholders. Participants discussed how the power structure of various communities needed to be understood because it determined which groups benefited from industry presence and which were marginalized, a key issue in social impact assessment. Community stakeholders vary across communities and could include political groups, property owners, and business owners. Identifying the political voice of a community is important in understanding the nature of the relationship between the community and the offshore industry.

Participants also identified minority groups or cultures in the region as stakeholders to be considered, such as Cajuns, the Houma, the Chitimacha, Vietnamese, and Hispanics. Some of these groups have specific relationships to the petroleum industry that may affect social institutions and culture.

Many of the important stakeholders that were identified interact with the petroleum industry through their use of the physical environment. These populations include those involved in fishing, shrimping, boating, diving, and other forms of recreation and commerce. Important commercial stakeholders may include those who use the GOMR ports and Gulf shipping lanes as a transportation corridor, as well as the GOMR ports themselves. In some areas, military uses are also involved.

Another important group discussed at length in all breakout sessions was the industry labor force. Participants wanted to know more about: the skills and educational requirements needed and how they've changed over time; the formal and informal educational systems that develop these skills; recruitment and retention issues; job quality; and the role of technological change on jobs.

Participants wanted to know about labor force differences across industry sectors. It was suggested that specific industry sectors may generate unique occupational communities or identifiable regional cultural differences. By understanding worker roles, positions, training, labor requirements, and stresses, we can better understand how industry trends may affect worker lifestyles.

Similarly, participants noted the importance of the socioeconomic networks of migrants working in the Gulf of Mexico region. Such networks help determine relationships to the industry, occupational niches, and the strength and ties to the local area as well as to their communities of origin. The partial integration and participation of migrants in the U.S. system may not be viewed as a negative impact by migrants themselves.

State and Local Variation

The third major crosscutting theme discussed across workshop groups was the importance of addressing the geographical variation in industry effects. A change in the industry may not only affect one place more than another, but it also may be a benefit to one place and a burden to another. Participants emphasized differences in oil-related outcomes among the Gulf states in topics as diverse as industry growth and capitalization; contributions to, and demands on, infrastructure and public services; and effects on the environment. Participants suggested that such state-level differences need to be documented and explained.

One critical factor noted by all the breakout groups was the difference in state fiscal structures. Because each community has different fiscal structures and sources of revenue, some workshop participants expected that this variation would be key to understanding socioeconomic effects related to the offshore industry. The fiscal structure information could indicate the level of economic dependency on various industry activities and sectors. On a related topic, the differences in community infrastructure are also directly tied to the fiscal policies of each locality and state. Although MMS has conducted a substantial number of community-level studies, much of the information is concentrated in a few local communities. The participants argued that we needed more information on fiscal structure and its effects across the region.

Workshop participants emphasized the importance of studying port communities in the GOMR. They wanted to know more about infrastructural demands and capacities of various ports, as well as their ability to adapt to industry change. Additionally, some participants wanted to research the potential socioeconomic effects resulting from the competition between Gulf ports, states, and foreign countries for OCS activities.

Understanding community networks across the GOMR was also identified as an important information need. Participants believed that identifying and comparing the type and scope of formal and informal networks would lead to a better definition of affected areas. Furthermore, these networks could explain why some communities vary in their adaptation to changes in oil and gas industry activities. Identifying affected cultural groups was also seen as important.

The identification and understanding of affected cultures or culture groups was also an important information need. Many of those listed are well known—the Houma, for example. Participants believed that networking methodologies might work for the Houma and some other groups, and particularly for newly in-migrated ones, but that generally cultures were too geographically and socially diffuse to apply this approach. In some areas, the populations of counties and parishes affected by particular industry sectors might serve as a reasonable substitute for "culture group." Participants emphasized the need to look at change over time and between generations. The MMS oil history project was seen as providing the kind of detail and historical depth that would be needed. The MMS should explore the possibility of dividing the Gulf coastal areas into something like "culture areas" to better address social and cultural change. Areas might be identified based on the sectors of the oil industry present, specific cultural entities present and local histories. The consideration of oil and culture should include the investigation of community identification with the industry. Impacted cultural groups were seen as immediately relevant to the topic of environmental justice. The development of a "mini-atlas" for the state of Louisiana of low income and minority populations was suggested. This could be expanded to cover the entire Gulf coast region. Studies of indigenous groups were suggested.

Context of Social and Environmental Change

A fourth common theme addressed was the importance of context in social and environmental change. The industry does not exist within a vacuum. Its effects must be considered within the context of other ongoing trends in the environment. Land loss, a critical problem in south Louisiana, was discussed as one example of this problem. Participants noted that the workshop was discussing the subtleties of oil industry effects on small-scale subsistence economies, social structural networks, life experiences of different generations, and such culture features as language, but that the movement of people in south Louisiana due to coastal erosion has more significant and immediate effects in all these areas.

Monitoring

The issue of monitoring socioeconomic effects of the OCS program arose in all the sessions. Long-term monitoring was discussed as a goal under NEPA and the OCSLA. In this regard, effective monitoring was identified as the best way to address the issue of cumulative effects. Monitoring was also discussed as a methodological tool or research strategy, one that might prove particularly useful in the context of the Gulf, a complex situation in which effects are ongoing, causal relationships are impossible to specify, and outcomes are difficult to predict. The practicality and usefulness of monitoring industry, communities, and social conditions were discussed. An essential component of effective monitoring is the development of a set of common indicators that could be collected and analyzed over a long time period and across the entire GOMR. Some participants suggested that the EIS be used as a methodological tool to evaluate community change.

Multi-Methods and Multiple Approaches to Issues

The last major crosscutting theme identified by the workgroups involved methodology. Workshop participants emphasized the need for various types of data (e.g., primary, secondary, cross-sectional, and longitudinal) and the application of a wide range of research methodologies and strategies (e.g., case studies, surveys, historical case studies, network analysis, ethnographies, and GIS). Overall, participants said it would be desirable to decide which data are most important to collect as well as to explore better methods of collecting it. Workgroups suggested that two valuable parts of this undertaking should be a more inclusive synthesis of research MMS has completed and, also, a more systematic use of information that MMS and other state and federal agencies already collect.

APPENDIX A

PARTICIPANTS

Full Name: Michael Robert Adamson, Ph.D.
Affiliation: Adamson Historical Consulting
Field of Specialty: Twentieth-Century U.S. History
Research interests and/or area of expertise: Business and labor history, with a particular interest in organizational behavior; economic development; urban history; international economic/financial relations; public policy history across these areas

Full Name: Diane Austin, Ph.D.
Affiliation: University of Arizona
Field of Specialty: Applied Anthropology
Research interests and/or area of expertise: Environmental anthropology; community-based research; risk perception; impact assessment

Full Name: William B. Bankston, Ph.D.
Affiliation: Louisiana State University
Field of Specialty: Sociology
Research interests and/or area of expertise: Crime/deviance; social change

Full Name: Vern Baxter, Ph.D.
Affiliation: University of New Orleans, Department of Sociology
Field of Specialty: Sociology
Research interests and/or area of expertise: Industrial organization and business structures

Full Name: John J. Beggs, Ph.D.
Affiliation: Louisiana State University
Field of Specialty: Sociology
Research interests and/or area of expertise: Stratification, social networks, labor markets, demography

Full Name: Mary Boatman, Ph.D.
Affiliation: Minerals Management Service
Field of Specialty: Chemical Oceanography
Research interests and/or area of expertise: Marine science

Full Name: C. Ray Brassieur, Ph.D.
Affiliation: Department of Sociology and Anthropology, University of Louisiana at Lafayette
Field of Specialty: Cultural Anthropology
Research interests and/or area of expertise: Louisiana folk culture; wetlands cultural ecology; traditional ecological knowledge

Full Name: Ralph B. Brown, Ph.D.
Affiliation: Brigham Young University
Field of Specialty: Rural Sociology
Research interests and/or area of expertise: Social and community impacts; social change and rural development

Full Name: Donald G. Callaway, Ph.D.
Affiliation: Senior Cultural Anthropologist, National Park Service, Alaska Regional Office
Field of Specialty: Cultural Anthropologist and Statistics
Research interests and/or area of expertise: Indigenous peoples of the far north with special emphasis on Alaska Natives; and indigenous peoples of the Russian Far East. Formal statistical analysis including social network analysis, and oral histories. Topical areas include globalization, socioeconomic and cultural change. Cooperative management regimes and traditional (local) ecological knowledge

Full Name: Joseph Christopher, Regional Supervisor, Leasing and Environment
Affiliation: Minerals Management Service

Full Name: Rodney Cluck, Ph.D.
Affiliation: Minerals Management Service
Field of Specialty: Sociology
Research interests and/or area of expertise: Environmental sociology; social change and development; race and culture; social impact assessment; energy

Full Name: Craig E. Colten, Ph.D.
Affiliation: Professor, Department of Geography and Anthropology, Louisiana State University
Field of Specialty: Environmental Historical Geography
Research interests and/or area of expertise: Urban pollution; hazardous waste; environmental justice

Full Name: Kimberley Cook, Ph.D.
Affiliation: Minerals Management Service
Field of Specialty: Anthropology
Research interests and/or area of expertise: Globalization and change; South America and Circum-Caribbean; aggression; women's studies

Full Name: Kristi A. R. Darby
Affiliation: Center for Energy Studies, Louisiana State University
Field of Specialty: Geology
Research interests and/or area of expertise: Upstream oil and gas, LNG

Full Name: Brian Diepold
Affiliation: URS Corporation and American University
Field of Specialty: Economics

Research interests and/or area of expertise: Industrial organization/applied microeconomics, with a focus on international mergers

Full Name: David Dismukes, Ph.D.
Affiliation: Louisiana Center for Energy Studies, Louisiana State University
Field of Specialty: Energy Economics, Policy/Technical Analysis
Research interests and/or area of expertise: Energy industries

Full Name: Craig J. Forsyth, Ph.D.
Affiliation: University of Louisiana, Lafayette
Field of Specialty: Sociology
Research interests and/or area of expertise: Social impacts of OCS activity; examining the impact at the family and community levels

Full Name: William R. Freudenburg, Ph.D.
Affiliation: Dehlsen Professor of Environmental Studies and Sociology, University of California, Santa Barbara
Field of Specialty: Sociology
Research interests and/or area of expertise: Resource-dependent communities/regions; technological controversies, risk, risk management, risk communication; environment-society relationships

Full Name: Peter H. Fricke, Ph.D.
Affiliation: NOAA/National Marine Fisheries Service, Office of Sustainable Fisheries
Field of Specialty: Sociology
Research interests and/or area of expertise: Fishing and fishery-related occupations; fishing- and maritime-dependent communities; coastal zone management; sea-use management; demography; ethnography and oral history of coastal residents; NEPA and marine law and policy

Full Name: Duane A. Gill, Ph.D.
Affiliation: Social Science Research Center, Mississippi State University
Field of Specialty: Sociology
Research interests and/or area of expertise: Environmental sociology, community, and risk/disasters (MMS OCS Scientific Committee Member)

Full Name: Robert Gramling, Ph.D.
Affiliation: University of Louisiana at Lafayette
Field of Specialty: Environmental Sociology/Social Impact Assessment
Research interests and/or area of expertise: Coastal resource development

Full Name: Scott A. Hemmerling
Affiliation: Doctoral Candidate, Department of Geography and Anthropology, Louisiana State University

Field of Specialty: Geography
Research interests and/or area of expertise: Environmental hazards; environmental justice; environmental philosophy; U.S. South

Full Name: Richard G. Hildreth, Ph.D.
Affiliation: University of Oregon Ocean and Coastal Law Center, Professor of Law and Director
Field of Specialty: Ocean Resources Management
Research interests and/or area of expertise: Sustainable use of living and non-living ocean resources

Full Name: Omowumi O. Iledare, Ph.D.
Affiliation: Center for Energy Studies, Louisiana State University
Field of Specialty: Petroleum Economics
Research interests and/or area of expertise: Oil and gas industry economics and policy analysis

Full Name: Jack Irion, Ph.D.
Affiliation: Minerals Management Service
Field of Specialty: Marine Archeology
Research interests and/or area of expertise: Deep-sea shipwrecks

Full Name: Mark J. Kaiser, Ph.D.
Affiliation: Center for Energy Studies, Louisiana State University
Field of Specialty: Energy Economics, Policy/Technical Analysis
Research interests and/or area of expertise: Decommissioning; fiscal systems; oil and gas infrastructure

Full Name: Gary Kane
Affiliation: The Kane Kompany, Inc. (President)
Field of Specialty: Offshore Oilfield Construction, Maintenance, and Inspection with an Emphasis on Projects Involving Divers and Remote Vehicles
Research interests and/or area of expertise: Human performance in the offshore environment, emphasis on safety and quality workmanship

Full Name: Maureen F. Kaplan, Ph.D.
Affiliation: Eastern Research Group, Inc.
Field of Specialty: Policy Analysis; Economic and Financial Analysis
Research interests and/or area of expertise: Policy analysis with a focus on the economic and financial impacts on industry from government regulations

Full Name: Constance C. Landry
Affiliation: Minerals Management Service
Field of Specialty: Administrative Assistant; Procurement Coordinator
Research interests and/or area of expertise: Louisiana cultures

Full Name: Shirley Laska, Ph.D.
Affiliation: Center for Hazards Assessment, Response and Technology; University of New Orleans
Field of Specialty: Sociology
Research interests and/or area of expertise: Environmental sociology/disaster management

Full Name: Dayna Bowker Lee, Ph.D.
Affiliation: Louisiana Regional Folklife Program, Northwestern State University
Field of Specialty: Cultural Anthropology
Research interests and/or area of expertise: Native American studies; Louisiana folklife; Creole studies; French Colonial Louisiana

Full Name: Asha Luthra
Affiliation: Minerals Management Service
Field of Specialty: Sociology
Research interests and/or area of expertise: Macro-criminology; social stratification; social impact assessment

Full Name: Harry Luton, Ph.D.
Affiliation: Minerals Management Service
Field of Specialty: Anthropology
Research interests and/or area of expertise: Social impact assessment

Full Name: Brian G. Marcks
Affiliation: Louisiana Department of Natural Resources, Coastal Management Division
Field of Specialty: Environmental Impact, Coastal Zone Management
Research interests and/or area of expertise: Wetland ecology; environmental impact evaluation

Full Name: Thomas R. McGuire, Ph.D.
Affiliation: Bureau of Applied Research in Anthropology, University of Arizona
Field of Specialty: Anthropology
Research interests and/or area of expertise: Natural resource-dependent communities; social impact assessment; public policy

Full Name: Wilbur E. Meneray, Ph.D.
Affiliation: Assistant Dean for Special Collections, Howard-Tilton Memorial Library, Joseph Merrick Jones Hall, Tulane University
Field of Specialty: Louisiana History
Research interests and/or area of expertise: Colonial Louisiana, Civil War, modern Louisiana politics

Full Name: Diana Davids Olien, Ph.D.
Affiliation: University of Texas of the Permian Basin
Field of Specialty: History of the American Petroleum Industry, with Emphasis on Texas

Research interests and/or area of expertise: History of the petroleum industry, American women's history

Full Name: Maida Owens, M.A.
Affiliation: Louisiana Folklife Program, Division of the Arts
Field of Specialty: Folklore/Anthropology
Research interests and/or area of expertise: Louisiana traditional cultures; public presentations; roadside memorials

Full Name: Frederick B. Palmer
Affiliation: Shell Exploration & Production Company
Field of Specialty: Government and External Affairs
Research interests and/or area of expertise: Providing public, social performance, reputation and local government relations support for Shell's operations in the United States.

Full Name: Michael E. Parker
Affiliation: ExxonMobil Production Company
Field of Specialty: Regulatory Advocacy
Research interests and/or area of expertise: Offshore operations and support; environmental compliance execution and monitoring; environmental impact assessment

Full Name: John S. Petterson, Ph.D.
Affiliation: President, Impact Assessment, Inc.
Field of Specialty: Applied Anthropology
Research interests and/or area of expertise: Technological disaster, community development, and social impact studies

Full Name: Tyler Priest, Ph.D.
Affiliation: History International, LLC
Field of Specialty: Business History
Research interests and/or area of expertise: History of the offshore oil industry

Full Name: Zeta R. Rosenberg
Affiliation: ICF Consulting LLC
Field of Specialty: Economist
Research interests and/or area of expertise: Oil markets

Full Name: J. Rhett Rushing, M.A.
Affiliation: Research Department, University of Texas, Institute of Texan Cultures at San Antonio
Research Interests and/or area of expertise: Folk narrative, folk religion, and Texan and folk food ways. Current projects include a study of curanderismo and stories of war.

Full Name: Edella Schlager, Ph.D.
Affiliation: School of Public Administration and Policy, University of Arizona

Field of Specialty: Natural Resources Policy
Research interests and/or area of expertise: Community-based governance of natural resources, particularly fisheries and water

Full Name: Joachim Singelmann, Ph.D.
Affiliation: Louisiana State University
Field of Specialty: Sociology
Research interests and/or area of expertise: Stratification, Demography, Rural Sociology

Full Name: Keith Storey, Ph.D.
Affiliation: Memorial University of Newfoundland, Department of Geography
Field of Specialty: Socioeconomic Impact Assessment
Research interests and/or area of expertise: Community-level assessments; impact management strategies; impact monitoring and auditing; human factors issues in offshore

Full Name: Kristen Strellec
Affiliation: Minerals Management Service
Field of Specialty: Economics
Research interests and/or area of expertise: Energy, environmental, and mineral economics; social impact assessment

Full Name: Charles M. Tolbert, II, Ph.D.
Affiliation: Baylor University
Field of Specialty: Sociology
Research interests and/or area of expertise: Socioeconomic and social demographic analysis

Full Name: Allen Verret
Affiliation: Executive Director, American Petroleum Institute, Offshore Operator Committee Consortium; Technical Advisor to the Deepstar CTR 6100 Regulatory Sub Committee
Field of Specialty: Petroleum Engineer
Research interests and/or area of expertise: Engineering management; construction; field operations; production and drilling; and work over operations; marine technology development; and oil spill response operations.

Full Name: Debra Vigil
Affiliation: Minerals Management Service
Field of Specialty: Environmental Studies
Research interests and/or area of expertise: Public relations and socioeconomics

Full Name: Barbara Wallace
Affiliation: U.S. Department of the Interior, Minerals Management Service
Field of Specialty: Environmental Studies
Research interests and/or area of expertise: Economic development

Full Name: Laura Renée Westbrook, Ph.D.

Affiliation: Director, Louisiana Regional Folklife Program at the University of New Orleans

Field of Specialty: English, Pubic Folklore

Research interests and/or area of expertise: Applied folklore and anthropology; public interpretation; culture and environment; culture and public planning.

Full Name: Dee Williams, Ph.D.

Affiliation: Minerals Management Service, Alaska Region, Environmental Studies Program

Field of Specialty: Anthropology; Political Ecology

Research interests and/or area of expertise: Social impact assessment, subsistence, traditional knowledge, environmental perception, common property resources

Full Name: Robert Winthrop, Ph.D.

Affiliation: Planning, Assessment, and Community Support Group, USDI Bureau of Land Management

Field of Specialty: Anthropology

Research interests and/or area of expertise: Participatory approaches to resource management; American Indian cultural rights; cultural conflict resolution

APPENDIX B

AGENDA

Tuesday, February 3, 2004

8:30–8:45	Welcome and Introduction. Mary Boatman
8:45–9:15	Social Impact Assessment and Offshore Oil and Gas in the Gulf of Mexico. Rodney Cluck
9:15–9:45	Legal Requirements for Social Impact Assessments under NEPA. Richard Hildreth
9:45–10:00	Break
10:00–10:20	The History of the Offshore Industry in the Gulf of Mexico. Tyler Priest
10:20–10:35	Directions for the Breakout Groups. Harry Luton
10:35–10:45	Brainstorming. Barbara Wallace
10:45–11:30	Organizing breakout groups.
11:30–1:00	Lunch
1:00–4:30	Breakout session 1
4:30–6:30	Reception

Wednesday, February 4, 2004

8:30–11:30	Breakout session 2
11:30–1:00	Lunch
1:00–4:30	Breakout session 3

Thursday, February 5, 2004

8:30–8:35	Introduction. Rodney Cluck
8:35–8:50	Onshore Social Impact of Oil and Gas Development from BLM. Robert Winthrop
8:50–9:20	Understanding Cultural Change in Alaska—Measuring of Federal Actions on Indigenous Communities. Donald Callaway
9:20–9:35	Group 1 Synthesis Presentation. Kristen Strellec
9:35–9:50	Group 2 Synthesis Presentation. Asha Luthra
9:50–10:05	Break
10:05–10:20	Group 3 Synthesis Presentation. H.B. Kimberley Cook
10:20–11:30	Synthesis discussion.
11:30 – 11:35	Closing remarks.

BREAKOUT GROUPS (second half of first day & second day)

APPENDIX C

PROPOSED STUDIES AND STUDY TOPICS

Workshop participants were encouraged to submit, in writing, study ideas that MMS could draw on for future studies planning. No attempt has been made to rank these suggestions. Instead, the submitted materials are provided in alphabetical order by title. A letter from the State of Louisiana providing study suggestions is attached to the end of this list.

(1) Project Title: Adaptive Strategies in a Changing Environment

Study Area: Mobile, Alabama to Brownsville, Texas with emphasis on the five Louisiana basins and the Atchafalaya basin above Morgan City. The five basins include the Lower Mississippi Valley, Terrebonne/Lafourche, Atchafalaya/Wax Lake, Teche/Vermillion, and Chenier.

Objectives: Plan to measure and describe how families and groups cope with fluctuations and impacts of the oil industry and changes in the social and natural environment. The areas of culture to be examined include family organization, knowledge of the environment, resource utilization, sense of place, social support networks of sharing, material culture, gender roles, language, community decision-making and political autonomy, oral history, cultural geography, ethnoscience, religious practices, intergenerational issues, associations and interest groups, and self identity and ethnic identity.

Methods/Study Design: ethnographic research, ethnohistorical research, survey, secondary sources, by informant, oral history, cultural geography, and ethnoscience.

Data Sources: community respondents, archives, photos, genealogies, civic records, public records, diaries, journals, U.S. Census, Bureau of Economic Research, land records.

Potential Challenges: scope, accessibility, sampling, non-response, and identifying cultural brokers.

Products: tapes/transcripts, community histories, ethnographies and community studies covering the above topics, SPSS data files (survey files), GIS mapping, photographs, and technical reports.

(2) Project Title: Business Event: Departure of Large Employer

Study Area: New Orleans, LA.

Proposed Period of Performance: Two years.

Objectives: Understand impact of moving management functions to Houston. Develop history of particular OCS-oriented business establishment. Study relationship of single establishment to community.

Methods/Study Design: Historical analysis, political analysis, engage civic leadership/business leadership, track the employer—did they really leave?

Data Sources: Corporate records, interviews, archived documents, newspapers, etc.

Potential Challenges: Access to data sources above; impact will be very diffuse due to urban area scale.

Products: report, papers/articles, possible book

Current Status of Information on this Topic (linkages): Not much, but one could draw on deindustrialization and plant closure literature.

Justification/Applicability of Information to Issues of Regional or Programmatic Concern: Key issue for New Orleans cultural identity/psyche. How involved are people in the community? Can we identify impact in large-scale urban areas?

(3) <u>Project Title</u>: Characterization of Support Activities in Gulf: Current Configuration and Potential Future Trends

Study Area: Gulf of Mexico.

Proposed Period of Performance: Six months. Might be scheduled to coincide with release of 2003 or 2004 data.

Objectives: To provide a financial and economic characterization of the support industries in the GOM OCS by project stage and project type.

Methods/Study Design: For every phase in a project, identify associated activities. Identify which activities are support activities and whether this designation might differ between independent and major oil companies. Identify companies participating in support industries, and use publicly available sources of information to identify company ownership, number of employees, revenues, use of contract labor, customer base and area serviced. Develop a series of matrices that summarize findings. Use the matrices to provide a qualitative analysis of the change in support activities as older, closer-to-shore projects end and fewer newer and further-from-shore projects come on line as a result of lease sales.

Data Sources: MMS publications and data, government data (Securities and Exchange Commission's EDGAR data base), financial data bases (e.g., Standard & Poor's, Dun & Bradstreet), company web sites, and newspaper articles.

Potential Challenges: Detailed information might not be available for smaller supporting industries. Businesses might not distinguish between operations in state waters and those in federal waters. Other information might involve purchasing data from a vendor such as InfoUSA or Dun and Bradstreet.

Products: The end-product is series of matrices that qualitatively characterize the GOM OCS support industries.

Current Status of Information on this Topic: This project would build on the information in *Deepwater Program: OCS-Related Infrastructure in the Gulf of Mexico: Fact Book.*

Justification/Applicability of Information to Issues of Regional or Programmatic Concern: First, a comparison of current support activities by distance from shore can indicate whether future development (e.g., deepwater areas) show different patterns, thus making estimates on socioeconomic impacts of future lease sales more accurate. Second, MMS can make a more informed estimate on the likelihood that support services will continue to come from established onshore areas compared to establishing new businesses in new areas. Third, MMS can indicate the number of small businesses currently supported by the offshore oil and gas activities in the GOM and, if possible, in the GOM OCS.

(4) <u>Project Title</u>: Closing of an Oil Business in an Urban Area

Study Area: Lafayette or any other city where an oil company moved out.

Objectives: Define the oil company (or a part of it) leaving the city. Examine the individuals and their families. What happens to them?

Methods/Study Design: Interviews.

Data Sources: Individuals/families.

Potential Challenges: None.

Products: Add to the literature on impact.

Current Status of Information on this Topic (linkages): Company movement has a dramatic effect—there may be no effect or less than previously thought.

Justification/Applicability of Information to Issues of Regional or Programmatic Concern: Oil companies move from cities, but do employees move out or not? Do families stay?

(5) Project Title: Community Response to Oil: Toward a Typology of Adaptation and Accommodation to Industry

Objectives: Case studies of various communities where extensive OCS activity exists. This will begin to identify the factors that mitigate effects.

Methods/Study Design: Develop a social historical view of each community. Over time, a typology of communities will develop. The typology will be along a continuum of variables.

Data Sources: Census data, other socioeconomic indicators, and ethnographic data.

Products: A typology of oil communities.

Current Status of Information on this Topic (linkages): Most research has concentrated on the boom/bust model.

Justification/Applicability of Information to Issues of Regional or Programmatic Concern: MMS is searching for a new paradigm to examine effects.

(6) Project Title: Competitiveness of Oil and Gas Service Industry in the GOM

Study Area: Industry Dynamics and Trends.

Proposed Period of Performance: January 2005–December 2007.

Objectives: Provides measures of the competitiveness of oil and gas field service industry in the GOM.

Methods/Study Design: Statistical and econometric analysis of data using indicators used in industrial organization.

Data Sources: Databases compiled by Minerals Management Service and processed by LSU Energy Studies for previous MMS funded projects.

Potential Challenges: Data collections and processing. Data continuity and availability. Accounting for mergers and acquisitions of firms within the service sector

Products: Final report, papers published in the proceedings of conferences organized by reputable professional organizations and refereed journal articles.

Justification/Applicability of Information: Assessing industry performance and cost of services in the GOM and its implications of on resource development in the OCS.

(7) Project Title: Description of Industry Impacts on Cultures by Historical Eras and Project Lifecycle Phases

Study Area: The geographic area of this project can potentially include the entire Gulf of Mexico Region. Study areas will be determined by the presence or absence of industry sector activity.

Proposed Period of Performance: This project can be carried at any level of complexity. It was suggested that a baseline study be carried out in which first era is described.

Description of Proposed Project: There are three proposed project eras. The first is an historical description of oil activities on land. The second era is an analysis industry activities carried out on the shelf. The third would be carried out in deepwater. The fourth would be an analysis of ultra-deepwater activities. These eras represent historical periods, beginning with industry activities on land and culminating in future activities in ultra-deepwater.

Project lifecycle phases would include descriptions of industry activities, resource needs, and impacts on culture. Particular or joint features of culture can be compared across eras for particular project phases. For example, what are the impacts of labor requirements on local communities in across time based on differential industry activities? After this information is collected, comparisons can be made between particular features of culture and between cultures as well.

Methods/Study Design: The initial project phase will be to conduct a study of initial onshore era. Ethnographic information and interviews can be obtained from industry experts from within the Minerals Management Service.

Data Sources: Ethnographic interviews, expert opinion, and historical sources.

(8) Project Title: Differences Among Independents and Major Oil Companies with Respect to Offshore Accident Rates

Study Area: Gulf of Mexico.

Proposed Period of Performance: Six months.

Objectives: Independents, both large and small, are playing an increasing role in the GOM OCS. The workshop identified the question of whether there is a difference in accident rates among the company types. That is, as independents play an increasing role, will there be a concomitant increase in accident rates?

Methods/Study Design: The researcher will investigate state and federal sources of injury and fatality statistics to identify the set of offshore oil and gas operations. Within the set, the researcher will work with the source to subdivide the observations into subsets based on company type (e.g., major, independent, contractor, service industry). If possible, the researcher will identify multiple years of data for the analysis.

The researcher will perform standard statistical tests (e.g., hypothesis testing, Chi-squire, trend analysis, etc.) to evaluate whether there are significant differences or trends in accident rates among different types of companies active in the GOM OCS.

Data Sources: MMS Accident Investigation data, Bureau of Labor Statistics (Fatal Occupational Injuries and Occupational Injuries and Illnesses), Occupational Safety and Health Administration data from Form 300 (formerly Form 200), Log of Work-Related Injuries And Illnesses and Form 301 (formerly Form 201), Injury and Illness Incident Report, and state data.

Potential Challenges: The potential challenges associated with this project are 1) confidentiality of the data and 2) separation of onshore and offshore activities.

Products: Identification of whether there is a change in accident rate among company types. If not, then changing role of independents in the GOM OCS is unlikely to lead to different (higher) injury rates.

Justification/Applicability of Information to Issues of Regional or Programmatic Concern: Independents are playing a growing role in the GOM OCS. A potential change in worker safety is one of the socioeconomic factors to consider when examining the changing nature of GOM OCS activities.

(9) Project Title: Doing Cumulative Effects Assessment

Study Area: National/International.

Objectives: Identification of "good practice" approach to cumulative effects assessment in context of GOMR (large scale industrial sectors; complex activity; wide-geographic spread).

Methods/Study Design: Literature review, focus groups (technical).

Potential Challenges: Few good examples of socioeconomic Cumulative Effects Assessment. Generally, fewer still for this type of problem.

Products: "Good practice" methodology proposal

Justification/Applicability of Information to Issues of Regional or Programmatic Concern: Multiple projects, programs have systemic, synergistic effects. Developing an appropriate approach to this meta-problem would provide guidance to MMS on impacts of past and future projects.

(10) Project Title: Environmental Justice at the Regional Level

Description of Proposed Project: To launch a study to investigate the longer term historical processes of settlement and population distributions along the Gulf Coast to consider a) whether regions other than Lafourche Parish are also likely candidates for EJ investigations; b) how the Houma came to be in the precise locations where they are now; c) whether current resettlement issues that arise from coastal subsidence can be framed and understood in a more EJ focused narrative; etc.

(11) Project Title: Environmental Justice from an Historical Perspective

Description of Proposed Project: To conduct an investigation of historical changes in community identification with the oil industry. This project would have environmental justice overtones but could also be framed well outside of environmental justice perspectives. The hypothesis might be that significant turning points in the industry correspond with notable and measurable changes in the way residents buy into and "own" the oil industry as an appropriate locus of regional identity and sense of place. For example, as the industry moves into deepwater and labor demands change, communities respond in measurable ways with greater distance from the industry. As erosion issues deepen, perhaps identifiable sectors of the community realign their cultural identities and associations with the industry. And it is possible that some broad scale "dispossession" or "redefinition" can be detected that would provide a vehicle to raise the hot political issues of renegotiating regional access to federal distributions of mineral royalties and/or compensation claims.

(12) Project Title: Examine the Role of Offshore Work in the "Saving of Culture"

Description of Proposed Project: 7 & 7 work allowed individuals to work in other areas during their off time—farming/shrimping. It provided extra money so they could shrimp and farm.

(13) Project Title: Exploration Efficiency of Firms of Different Sizes: Implications on GOM Petroleum Resource Development

Study Area: Industry Dynamics and Trends.

Proposed Period of Performance: October 2005–September 2007.

Objectives: Identify and model measures of exploration efficiency by firms of different sizes to evaluate the trends in OCS industry dynamics and strength within the context of global petroleum supply and pricing.

Methods/Study Design: Collect and process OCS data on exploration and production activity on in the GOM. Develop models to evaluate E&P activity on the GOM OCS.

Data Sources: Databases compiled by Minerals Management Service on drilling, production and reserves and processed by LSU Energy Studies for previous MMS funded projects.

Potential Challenges: Data collections and processing. Data continuity and availability. Developing algorithm to manage and process data into a useful format and to allocate reserves on lease basis to firms of different sizes.

Products: Final Report, papers published in the proceedings of conferences organized by reputable professional organizations and refereed journal articles.

Current Status of Information on this Topic: Update past studies conducted by researchers at LSU Energy Studies on the expanding role of independent oil and gas operators and profitability of OCS leases.

Justification/Applicability of Information: Provide measures or indicators of industry trends and dynamics to quantify future GOM activity and output.

(14) Project Title: Factors Affecting Future Petroleum Exploration Productivity in the GOM

Study Area: Industry Dynamics and Trends.

Proposed Period of Performance: January 2005–December 2007.

Objectives: Identify factors and issues that may affect future OCS oil and gas productivity and the implications of such issues and factors on petroleum resource development in the GOM.

Methods/Study Design: Statistical and econometric analysis of data on the determinants and output of E&P activity on the GOM OCS.

Data Sources: Databases compiled by Minerals Management Service and processed by LSU Energy Studies for previous MMS funded projects.

Potential Challenges: Data collections and processing. Data continuity and availability. Developing algorithm to manage and process data into a useful format.

Products: Final report, papers published in the proceedings of conferences organized by reputable professional organizations and refereed journal articles.

Justification/Applicability of Information: The Gulf of Mexico OCS is expected to continue to play a major role in the U.S. efforts to import domestic production and reduce oil import. Identifying the factors affecting OCS exploratory productivity and trends in petroleum supply would facilitate the understanding of the potential of the Gulf in this effort.

(15) <u>Project Title</u>: Follow-up Study to Current MMS Study Entitled: Environmental Justice Considerations in Lafourche Parish, Louisiana (OCS Study MMS 2003-038)

Study Area: Lafourche Parish.

Description of Proposed Project: This study will be a follow-up to the environmental justice study just concluded by Scott Hemmerling, et. al. centered around Lafourche Parish. This study found that there is evidence that minority groups (Houma Indians) are disproportionately affected by OCS planning activities. Based on this finding this study would ask the following questions: 1) What are the consequences of that disproportionate exposure and risk of exposure? 2) Can any specific negative impacts be documented, or even shown to be plausible? and 3) How do Houma traditional activities interact with the "nuisance" in ways that either diminish or enhance the disproportionate risk?

(16) <u>Project Title</u>: Historical Examination of the Offshore Drilling Industry Impact on the Culture Groups of the Gulf of Mexico Region

Study Area: GOMR—Florida Panhandle continuing West and South to Brownsville, Texas

Proposed Period of Performance: Immediately preceding any other studies in the region (for purposes of historical accuracy, contextual authenticity, and troubleshooting).

Description of Proposed Project: Objectives: To document and articulate the impact of the offshore drilling industry on coastal culture groups historically.

Methods/Study Design:
1. Phase One—This project will involve an extensive archival and literature search of historical interactions between the offshore drilling industry and coastal groups—from the beginnings of the land-based drilling industry through the move offshore and to the present. This should take place in the archival holdings and collections of facilities already known to focus on the oil industry and its service industries.

2. Phase Two—This project will require oral history, life story, and ethnographic interviews with persons involved in and affected by all manners of the offshore drilling industry. Beginning with community leaders, retired employees, and other primary sources, the project will spread outward to encompass persons more peripherally connected to/impacted by the industry.

This project will deal with human stories and artifacts and will draw its conclusions from the ground up. Emphasis will be placed upon each narrator's evaluations and not a predetermined range of acceptable responses that generate statistical models.

Data Sources: Primary sources will be employees and former employees of oil industry and service industry entities. Focus here will be on not only the employees, but on their families. Secondary interviews will be with those less directly associated with the industries— infrastructure, housing, local government, schools, chambers of commerce, retail outlets, churches, etc. We are looking for patterns in the industry as they are perceived by the very people they impact.

Potential Challenges: This is relatively new territory for those dependent upon charts and graphs to guide their thinking. We feel that the failing of many statistical models is that they do not "listen." Instead they provide a range of answers and ask subjects to respond within a given framework. This model allows the actual human being to put their own feelings into their own words and we propose to find patterns in their responses from which we will base our conclusions.

Potential challenges will include community attitudes towards the oil industry, the federal government, and perhaps outsiders in general. Also, personal experiences, relationships, and histories with the industry.

Oral histories and life story interviews are incredibly powerful and meaningful largely because they take place face to face with real human beings (not survey forms) and they tend to validate the informant's life experiences. With this is mind, there must be a component built into this study to train interviewers in techniques and technical documentation skills.

Ideally this study would be multi-faceted and coordinated across the GOMR by several institutions already capable of instructing interviewers in the methods and mechanics of oral history collection. This too, would be a challenge.

(17) Project Title: Impacts of Change on Other Resources Used by Local Populations and Communities

Study Area: Coastal parishes and counties.

Proposed Period of Performance: Two years, update every five years.

Objectives: Identify, describe, and value (social and economic) resources used by local people and communities. The resources may be natural resources, jobs, culture, etc. Identify the impacts of oil industry operations on these resources and assess the outcomes under difference change agent scenarios.

Methods/Study Design: Ethnographies of local populations and communities, map resources and areas of resource use, analyze impacts (e.g., effects on fishing and hunting by dredging new channels).

Data Sources: Census, land use maps and plans, employment data, oil/industry plans and proposals, historical records.

Potential Challenges: None.

Products: Materials for EIS, EJ, and cumulative impact assessment; understanding of socio/cultural issues related to proposals.

Current Status of Information on this Topic (linkages): None in Gulf States—some in Alaska.

(18) Project Title: Impacts of Offshore Development on Industry Migration and the Managerial/Administrative Workforce in the Gulf Coast Region

Study Area: Houston, New Orleans (and, possibly, Lake Charles, Lafayette, Corpus Christi, and Beaumont/Port Arthur).

Proposed Period of Performance: 2005–2006.

Objectives: To examine the impact of offshore development in the Gulf of Mexico on the patterns of industry migration to and within the Gulf Coast, focusing on the administrative and managerial workforce. The idea is to get a better understanding of the spread effects of the growth and development of the offshore industry beyond worker communities along the coast.

Methods/Study Design: The method would be historical research combining archives, statistical data and oral history interviews. The project would take as a model William Cronon's *Nature's Metropolis: Chicago and the Great West,* which reveals the environmental, ecological, and economic linkages between Chicago and its Great West hinterland in the late-nineteenth century. This study would seek to provide a similar analysis of Houston and New Orleans for the post-World War II period, examining the changing role of these two cities as the managerial hubs of the Gulf Coast offshore industry, and the impact of offshore oil and gas on the urban and regional geography of business and white-collar employment.

Data Sources: Source would include oral history interviews, statistical and census data on housing, employment, migration, etc.; accessible company information and publications; archival research at the City Archives and Special Collections at the New Orleans Public Library, the Houston Metropolitan Research Center, the Harris County Archives, and information from other relevant archives and libraries.

Potential Challenges: The main problem will be limiting the scope of the study and isolating the historical impact of the offshore petroleum industry from larger oil industry impacts on the region.

Products: A study as described in "Methods/Study Design."

Current Status of Information on this Topic (linkages): Very little has been written on the role of the oil and gas industry in the modern history of New Orleans and Houston. Joe R. Feagin's *Free Enterprise City: Houston in Political and Economic Perspective* (1988) is the only detailed study of business and urban development in Houston, but it does not address the issue of offshore oil or examine the growth of Houston in light of Gulf Coast regional development. Craig Colten's edited collection, *Transforming New Orleans and Its Environs* (2000) discusses industrial pollution and the petrochemical industry along the lower Mississippi River industrial corridor, but does not connect New Orleans to the "upstream" segments of the oil industry, including offshore Gulf of Mexico.

Justification/Applicability of Information to Issues of Regional or Programmatic Concern: This study idea was discussed and featured both in the study group on "Industry Trends and Dynamics in the Gulf of Mexico Region" and the one on "Community-Level Impacts of Oil and Gas Activity in the Gulf of Mexico Region" at the Social and Economic Planning Workshop, February 3–5, 2004. It will address a broad consensus of the workshop that we need more empirical and historical (or cumulative) approaches to understanding regional-level impacts of the offshore oil industry in the Gulf. In their analysis of social impact assessment, Harry Luton and Rodney Cluck point out that in the Gulf of Mexico, "Most impacts are not rural; most impacted communities are not physically or culturally isolated; most impacted economies are not agricultural; and most impacts and not caused by demographic change." This study will attempt to identify some of the urban, industrial and geographically integrated impacts of offshore development in the Gulf over time.

(19) <u>Project Title</u>: Levels of Action/Autonomy in the Interfacing with the Oil Industry

Study Area: A two-state or so region of the Gulf comprised of four to six community cases.

Objectives: Document which actors initiated or otherwise moved that a community get x or y industry, plant, etc. associated with the oil industry. Do communities actively seek certain plants, etc. to locate them or are they targeted by the county, state, industry? It may address the variables common across those communities that seem to better control their own destinies while using the oil industry at the same time to get them.

Methods/Study Design: Specifically trace the origins of certain plants, headquarters, etc. by documents and interviews in each town. Have a range of towns based on their ability to retain population, etc.

Data Sources: See above.

Potential Challenges: Finding and accessing key players at the state and industry levels.

Products: An analysis showing at what thresholds communities maintain their ability to act while still interfacing with industry and the state.

Current Status of Information on this Topic (linkages): I have published two studies that have used this approach: one on the Mercedes Bay plant locating in Alabama and the other on how a small town in Missouri established its industrial based in the 1950s long before other small towns did so.

Justification/Applicability of Information to Issues of Regional or Programmatic Concern: It would show levels of integration across state-community-industry, etc. and the points where each has more or less autonomy to act. This would help communities in particular know better where they can better leverage their ability to act to secure their best deals, while still allowing for the company to get what it wants.

(20) Project Title: Modeling the Impacts of Louisiana Coastal Land Loss on OCS Exploration and Production Activity

Study Area: Central Gulf.

Proposed Period of Performance: Two years.

Background: Coastal Louisiana is experiencing persistent land loss estimated at 25 square miles/year. Much of the infrastructure that supports Gulf OCS activity (pipelines, transportation facilities, preliminary processing, etc.) lies in these coastal parishes.

Objectives:
1. Identify current and anticipated risk due to land loss to the infrastructure and OCS support sectors to which they belong.
2. Examine the ways in which this impact would affect the bottom-line costs for Gulf OCS exploration/production.
3. Consider how the land loss may shift infrastructure and support sector activities to other Gulf locations as a result.

Methods/Study Design:
Team: Coastal geologists, environmental sociologists, natural resource economists.

Modeling land loss, identifying infrastructure and support sectors in shadow of impact. Estimating costs of damage, reconstruction, retrofitting for strengthening through interviews with industry technical representatives. Through scenario development of impacts, interview industry representatives for assessments of business decisions. Collect information from any sector companies that have already made decisions to move infrastructure and business activity due to land loss.

Data Sources: U.S.G.S., N.O.A.A., U.S. Army Corps of Engineers, MMS, U.S. Census County Business Patterns, industry organizations, businesses, representatives of businesses through in-depth focused interviews.

Potential Challenges: The potential challenges will be predicting risk as outcome of land loss. In addition, prediction of future business changes due to the risk is also a prediction rather than a measurement of past actual occurrence.

Products: Reports disseminated by MMS, refereed publications, presentations at industry and academic conferences.

Current Status of Information on this Topic (linkages): There have been no studies to our knowledge on this topic. Social/economic impact assessments in general of Louisiana coastal land loss are beginning to be undertaken.

Justification / Applicability of Information to Issues of Regional or Programmatic Concern: The findings will be useful for better appreciation of the business and community challenges that coastal Louisiana will face with land loss with regard to OCS activity. Equally useful will be the findings for an appreciation of the impact of coastal land loss on the revenue stream to the federal government from OCS production.

(21) Project Title: Monitoring Socioeconomic Impacts

Objectives: Attempt to define objectives of project/program/etc. Monitor whether these objectives have/have not met these objectives; determine whether or not management methods were effective in achieving their objectives.

Methods/Study Design: Secondary sources review, interviews, technical focus groups.

Potential Challenges: Environmental monitoring practices well-established focus-economic monitoring not so. Issues of objectives of monitoring, monitoring indicator choice, monitoring responsibility and commitment are poorly developed.

Products: Proposed methodology for how to monitor effects.

Justification/Applicability of Information to Issues of Regional or Programmatic Concern: Follow-up of EIS findings, proposed management strategies is essential if there is to be feedback and any "learning from experience". This information can be applied to future EA studies. Enables identification and promotion of both positive and negative outcomes. Monitoring programs designed to measure whether you achieve "what was wanted" not "what was expected."

(22) <u>Project Title</u>: OCS-Oil/Gas Employment as Part of Household Economic Strategies

Objectives: The socioeconomic impacts of closing OCS oil and gas-related facilities depends on the place of such employment in household survival strategies. This study will examine a sample of current and displaced oil/gas workers to identify a range of such
Survival/lifestyle strategies and to assess the effectiveness of such strategies in mitigating job displacement. Such strategies may include fall-back or job-subsistence living; periodic migration to other communities for oil/gas-related work; shifting to other work with compatible skills.

(23) <u>Project Title</u>: Oil and Community Networks

Study Area: Selected communities on the GOMR.

Objectives: Nature and extent to which oil-related activities influence community character and organization.

Methods/Study Design: Survey/interviews/review secondary sources.

Justification/Applicability of Information to Issues of Regional or Programmatic Concern: Oil and gas activities potentially have significant economic impacts at the community-level. The purpose would be to trace the social and political mechanisms that underlie and help in the organization of these activities.
- direct/indirect economic effects
- evolution of support services/technologies/capabilities
- ways in which oil and gas-related sectors are treated by local authorities (taxation, planning, etc.)
- involvement of industry within community/donations, political positions held, service organization involvement.

(24) Project Title: Oil and Gas and Houston Neighborhoods

Study Area: Houston Census Tracts.

Proposed Period of Performance: Three years.

Objectives: Identify industrial tracts of oil and gas workers in Houston. Profile demographics of oil and gas workers/neighborhoods. Associate temporal shifts in neighborhood patterns with key events in industry/geopolitics.

Methods/Study Design: Use decennial household Census data, explore new American Community Study as source for move frequency information, and standard demographic shift-share analysis.

Data Sources: Internal long form decennial Census data (1970, 1980, 1990, 2000)
American Community Study (1996, 1998, 2000, 2002, 2003, …)
Local data from planning agencies, etc.

Potential Challenges: Six months to secure access to data; 1970 Census data may not be Permeable in the time frame specified; local data is sketchy before 1990.

Products: Thesis/dissertation by graduate student, MMS report, academic articles, maps, tabular material subject to Census disclosure.

Current Status of Information on this Topic (linkages): No study at this level of detail in large-scale urban area.

Justification/Applicability of Information to Issues of Regional or Programmatic Concern: Emphasis on large urban area, detailed description of oil and gas workers and families, coverage of administrative and production workers, and exploration of new data source (ACS) which will replace the Census long form in 2010.

(25) Project Title: Preliminary Characterization of the Socioeconomic Impacts of a "No Action" Scenario

Study Area: Gulf of Mexico.

Proposed Period of Performance: Two years.

Objectives: Typically, a discussion of a "no action" scenario in an environmental impact statement is described in terms such as "all impacts, positive and negative, associated with the proposed actions would not occur." The infrastructure and employment in the GOM OCS and supporting on-shore communities, however, reflect a half-century of oil and gas development.

58

Oil and gas are non-renewable resources that, in aggregate, show an annual decline. An evaluation of a "no action" scenario should reflect the 1) absence of work for the support industries serving the exploration, delineation, and platform fabrication phases; 2) ensuing shift in ownership in existing structures; 3) declining need for support services during the production period; and 4) increase in the need for decommissioning support services. This project focuses on items 2 through 4.

Methods/Study Design: The goal is to develop a simplified tool for evaluating the impacts of a "no action" scenario. For current, existing projects in the GULF OCS, the steps would be: 1) develop a snapshot of current operations in the GOM OCS; 2) classify existing projects into model projects; 3) develop a financial model for model project; 4) record numbers and types of projects estimated to go out of production each year; 5) evaluate loss in production phase support services; and 6) evaluate increase in decommissioning phase support services. The financial model would incorporate logic to evaluate when project ownership might change (e.g., from major to large independent to small independent) due to different cost structures.

Support industry activities do not necessarily have a linear relationship with project production levels. It takes the same amount of activity to transport supplies a given distance from shore/port whether the project is at full production or nearing the end of its economical lifetime. The next step is to integrate the findings about the number of uneconomical projects with the matrix of support activities to qualitatively estimate the loss in production support services and the increase in decommissioning/ demolition support services over time.

Data Sources: MMS data bases supplemented as necessary by API data on ownership and operating costs. The models are intended to be flexible and provide an indication of possible future scenarios, rather than an accurate modeling of every project.

Potential Challenges: Adjusting the balance between broad brush assumptions and detailed data to provide a flexible yet accurate projection of possible future scenarios.

Products: An estimate, perhaps qualitative, of the decline in the need for support services in onshore communities in the absence of future lease sales.

It is possible that the model might identify a situation where it is the bulk of the smaller, nearing-the-end-of-life projects that require the majority of support services while the majority of production is coming from a small proportion of giant producers. Such a situation might lead to negative impacts on the onshore communities that support production phase activities. That is, a "no action" scenario is not simply the absence of the positive and negative impacts associated with a lease sale.

Justification/Applicability of Information to Issues of Regional or Programmatic Concern: Each EIS contains an evaluation of a "no action" scenario. This project would provide a beginning basis to evaluate the negative impacts of a "no action" scenario. Benefits of a lease sale are measured from the "no action" baseline and MMS might decide to present the alternative perspective where additional lease sales maintain (or prevent the erosion of) current socioeconomic conditions in the onshore communities. The study might indicate areas of a declining need for support activities in one region that might be available to support activities in another part of the Gulf of Mexico and thus alleviate the pressure for new onshore development.

(26) Project Title: Role of Elites in Community Response to Oil Activity

Study Area: Small city in Louisiana.

Proposed Period of Performance: 1970–2004.

Objectives: Elites are a valued community resource. Their existence determines the survival of a community. This project will study whether community elites are old-timers who stay or are newcomers who leave. A large part determines the effect of an oil downturn.

Data Sources: Interviews with community leaders.

Products: Develop theories of old-timers v. newcomers.

Current Status of Information on this Topic (linkages): Role of elites in community development exists.

Justification/Applicability of Information to Issues of Regional or Programmatic Concern: Learn the role of leadership—mitigating effect on community.

(27) Project Title: Shell Shock

Study Area: New Orleans, LA.

Proposed Period of Performance: 2005–2006.

Objectives: Community and firm-level study that aims to demonstrate impacts that quantitative data at the metropolitan level conceals. Do a study of how the arrival and consequent leaving of Shell corporate headquarters in New Orleans. It is bounded by time with Shell locating in the Gulf/New Orleans at the time of offshore drilling and them leaving in 2002 or so. How did this departure affect community identity, local politics, shifts in the tax base, shifts in civic engagement, etc.? Characterize Shell corporate presence—expectations. Effects on robustness of civic groups, tax base, charitable contributions, employee wages, social capital, and perceptions of the community. Characterize the activities of employees.

Methods/Study Design: Historical using both archives and interviews with key persons also company records and extant data (Census, etc.) on tax bases, newspaper accounts. Examining the managerial decision-making that determined levels of staffing in New Orleans and ultimately the closing of the headquarters.

Data Sources: Newspapers, company publications, trade publications, interviews, other extant data sets, New Orleans publications/magazines, histories of Shell, in-house Shell publications, and oral histories.

Potential Challenges: Access to some sources of data (people not willing to talk). No access to company sources and data/information, etc.

Products: A history of the effects that one major oil company has on a city when it moves in and out.

Current Status of Information on this Topic (linkages): Information can be found in the trade press on Shell. There is also information about the shutdown of oil businesses and infrastructure, especially for the bust following the collapse of oil prices in 1986.

Justification/Applicability of Information to Issues of Regional or Programmatic Concern: It would give a better picture of what potentially happens at the urban level as a result of a very specific segment of the Gulf oil industry—most specifically, with offshore oil drilling.

(28) Project Title: Social Networks and Social Capital in Oil-Related Communities in the GOMR

Study Area: Selected Communities along the Louisiana and Texas Coast.

Proposed Period of Performance: 2005–2008

Objectives: Develop an understanding of social networks and social capital of residents of oil-related communities. To what extent do these social networks coincide with the geographical/political communities? How does the social capital embedded in these networks facilitate the actions of individuals within these communities? Are the networks and social capital of oil/gas involved communities different from those not in oil/gas-related activity?

Methods/Study Design: Guided interviews to gather information from individuals in these communities.

Data Sources: Original data collection—public data sources to define oil/gas-related communities.

Potential Challenges: Identifying and/or getting access to oil and gas-involved individuals.

Products: Data set from interviews. Report of findings from analysis of differences between oil and gas and non-oil and gas involved and the implications for the ability to cope with oil and gas activity in area.

Current Status of Information on this Topic (linkages): Information on spatial communities but not for social networks and social capital.

Justification/Applicability of Information to Issues of Regional or Programmatic Concern: Social networks and social capital are important factors in how communities and individuals can cope with economic activity. Understand how social communities (defined by networks?) conform to political/geographic bounded communities.

(29) Project Title: Structural Trends in The Global Petroleum Industry as They Impact the Gulf Of Mexico Region

Study Area: Socioeconomic Research Issues.

Proposed Period of Performance: start either June or September, 2004 and finish either in December 2005 or May 2006.

Objectives: Explore industry and corporate governance and structure as they impact Central Gulf of Mexico (GOM) offshore oil investment and employment outcomes. Locate viability of GOM in dynamics of global petroleum industry. Specifically identify key players offshore and document the effects of joint ventures, mergers, larger companies operating in deepwater, foreign state oil companies, and key government regulatory/incentive policies on offshore investment and employment.

Methods/Study Design: Trend analysis, largely using existing and available data on the strategies and financial operation of 40-50 principal companies operating in the Gulf. Proposal is to update and analyze an existing data base built by the author (see V. Baxter, Chapters 2-3 in "Impact of Offshore Oil Exploration and Production on the Social Institutions of Coastal Louisiana," MMS 93-0007).

Financial data will be updated on about the largest 40 private companies and 10 or so foreign state owned companies active in the GOM. Data will be collected on joint ventures, mergers, number of local exploration and production offices, offshore lease purchases, wells drilled, employment, and other important variables related to offshore activities.

Data Sources: MMS data, including but not limited to annual OCS Statistical Summaries, economic forecasting data collected by ICF Consulting study of offshore expenditures, etc. Industry publications, such as Oil and Gas Journal, Pennwell's USA Oil Industry Directory, Pipelines and Facilities Database, and other publications.

Company financial data, available through such sources as Moody's Industrial Manual and Compact D/SEC online data base.

Personal interviews with industry sources as necessary.

Potential Challenges: Locating some data may be challenging, but most sources are secure and reliable.

Products: Report to MMS that can include raw data as well as summary and analysis: at least one conference paper; ultimately, a book manuscript.

Current Status of Information on this Topic (linkages): The proposed study is largely an update of a previous study done by the author (See MMS 93-0007) ten years ago. So far as I know, no comparable study of oil industry structure and trends is available that specifically refers to investment and employment outcomes in the offshore GOM. Scholarly research is plentiful on the governance and re-structuring of the petroleum industry, but this study places a unique focus on the offshore GOM.

Justification/Applicability of Information to Issues of Regional or Programmatic Concern: Study of industry trends, company strategies and structures is invaluable organizational analysis to complement economic analysis of the impacts of offshore oil activity. Outcomes of interest include investment and employment impact of the offshore oil industry in the GOM as they impact many other outcomes such as housing markets and regional economic development.

(30) Project Title: Untitled

Proposed Period of Performance: Three months.

Objectives: Develop a method and proposal for the study of impacts of OCS oil/gas industry on urban areas of GOMR.

Methods/Study Design: Examine several case studies of business closing/relocation as this affected a metro area (preferably in same sector, e.g., auto industry). Identify most relevant methods and types of data. Use these in OCS oil an gas proposal.

Products: Literature review, evaluation of several case studies, and proposed methods and data for OCS study.

(31) Project Title: Urban Communities and Infrastructure Impacts

Study Area: Large Urban Area (population > 50,000).

Objectives: Identify neighborhoods and populations likely to be impacted by changes in the oil industry's operations. Describe existing urban infrastructure and identify shortfalls and areas of excellence. Describe and quantify cumulative and proposed changes in the oil industry.

Methods/Study Design: Conduct neighborhood ethnographies; demographic overview of urban area; comprehensive review and valuation (social and economic) of urban infrastructure identifying critical variables/measurement points; evaluate neighborhoods and minority and/or poor populations living areas using infrastructure measurement points; evaluate cumulative and proposed changes using infrastructure measurement points; undertake impact analysis.

Data Sources: Census; urban area government; federal and state agencies; business plans and proposals of oil industry; fieldwork and ethnographies.

Potential Challenges: Obtaining clear reference and measurement points (variables)
Products: Description of human environment (EIS), description and analysis of impacts on special populations (EIS – EJ), description of cumulative impacts on neighborhood and urban area infrastructure, assessment of impacts under different change scenarios, and data bank.

Current Status of Information on this Topic (linkages): Census, population, and infrastructure data generally available but not brought together. Neighborhood and special population ethnographies not normally available.

Justification/Applicability of Information to Issues of Regional or Programmatic Concern: Develop baseline and analysis for use in EIS and assessment by MMS, industry, and local government.

(32) Project Title: Who Uses Offshore EIS Documents?

Study Area: GOMR.

Objectives: To determine who uses the findings/data from MMS EIS documents and for what purposes.

Methods/Study Design: Survey/interviews

Products: Report, recommendations on format/information to be included in EIS.

Current Status of Information on this Topic (linkages): Anecdotal.

Justification/Applicability of Information to Issues of Regional or Programmatic Concern: EIS documents are required. Determining who uses them and for what purpose could improve their utility as planning documents to promote/prevent/manage offshore development.

(33) <u>Project Title</u>: Work and Lifestyle

Study Area: GOMR.

Objectives: Examination of degree to which offshore worker defines or is defined by the worker's lifestyle and activity patterns.

Methods/Study Design: Survey of offshore workers; interviews. Determination of extent to which workers are 1) solely oil workers and 2) have secondary/other employment.

Products: Report outlining extent to which there are oil/non-oil activity linkages; implications of such linkages.

Current Status of Information on this Topic (linkages): Very limited.

Justification/Applicability of Information to Issues of Regional or Programmatic Concern: Workers with secondary employment/interests may be less affected by a downturn in the offshore oil and gas sector. Some may be able to return to commercial shrimping, fishing, etc. Rotational work in the offshore industry may support lifestyles that other work systems cannot. This may lead to higher levels of satisfaction among the workforce. In turn, this may have implications for worker turnover, productivity, worker health, etc.

State of Louisiana

KATHLEEN BABINEAUX BLANCO
GOVERNOR

DEPARTMENT OF NATURAL RESOURCES

March 8, 2003

Asha D. Luthra
Sociologist
U. S. Department of the Interior
Minerals Management Service
1201 Elmwood Park Blvd.
New Orleans, LA 70123

Re: MMS Request for Socioeconomic Research needs to improve Environmental Impact
Statements (EISs), Gulf of Mexico Outer Continental Shelf (OCS) Lease Sales

Dear Ms. Luthra:

In response to your February 11, 2004 e-mail request to provide study ideas as a followup to the February
3-5 Workshop, we have discussed in-house and come up with the following suggestions for your
consideration in developing future socioeconomic studies related to Gulf of Mexico Lease Sale EISs.

Review of MMS documents over several years indicates that there is no evident means or feedback
mechanism in place for MMS to review the data sources and predictive models used to estimate the
environmental and socioeconomic impacts presented in EISs for OCS Lease Sales. In short, it is not
clear that MMS provides validation that those data and models are accurate and current. State and
federal agencies rely heavily on these EISes in making decisions concerning environmental,
socioeconomic and infrastructure impacts on the Louisiana Coastal Zone. Because of the time lag
between EIS preparation and the subsequent development of OCS leases described in the document,
the effects of multiple OCS Lease Sales overlap and produce a confusing picture of impact cause and
effect. In order to assess the reliability of past and future EISs, a comprehensive assessment is needed
to compare actual development and impacts resulting from particular lease sales, with the predictions in
their respective EISs. The objective is to fine-tune predictions for future lease sales and so produce a
more reliable planning tool.

Another potentially useful socioeconomic study would be an industry survey to determine the effects of
the rapidly expanding liquified natural gas (LNG) import industry on oil and gas exploration in the Gulf
of Mexico OCS. We suggest examining the industry's changing needs for contractors and support
facilities, employment levels and trends, employee skill and education requirements, and the adequacy

COASTAL MANAGEMENT DIVISION   P. O. BOX 44487   BATON ROUGE, LOUISIANA 70804-4487
TELEPHONE (225) 342-7591   FAX (225) 342-9439   www.dnr.state.la.us
AN EQUAL OPPORTUNITY EMPLOYER

of existing port facilities and infrastructure such as navigation channels and transmission pipelines.

A third study of interest to Louisiana would look into the potential socioeconomic effects resulting from the competition between ports and shipyards that support OCS activities and offshore module fabrication. The continued maintenance and deepening of channels to various ports often is justified on the basis of jobs which will be maintained or created as a result of OCS and other development activities. A deeper channel, the argument goes, allows local shipyards to bid on more and bigger construction projects or fabrication facilities. This enhanced capability is projected to result in improvements to local and state tax revenue and economic development. However, the assumption appears to be that these economic development and employment opportunities are either "newly created" or are gained from or lost at the expense of other states or countries.

For the purposes of state planning, it may or may not make a significant difference whether an OCS construction project was awarded to Port "A" versus Port "B", if they were both in-state. However, if Port "B" was located in another state or country, there may be strong reasons for the state to support that port enhancement.

Questions to be investigated: (1) With whom do Louisiana companies and ports involved in the construction of offshore modules compete? Each other? Nearby states? Other US ports? Foreign facilities? How successful are Louisiana ports?
(2) How does winning/losing a contract at a particular facility translate into jobs and economic benefits, or changes in the "structure" of employment, within the state?
(3) What impacts to the state result from the deepening of a port, such as is being proposed for the Port of Iberia, the Atchafalaya River, the Houma Navigation Canal, etc.? What is the impact to the state if some or all of these channels are not deepened?
(4) Would any of these facilities really be positioned for bidding on fabrication contracts that will be used elsewhere, even with a larger channel?
(5) How large is the demand for construction facilities for offshore modules, in the Gulf of Mexico and worldwide?

If you have any questions with regards this matter, please contact me at 225-342-7939 or 1-800-267-4019.

Sincerely,

Brian G. Marcks
Consistency Analyst

APPENDIX D1

LEGAL FRAMEWORK FOR MINERALS MANAGEMENT SERVICE SOCIOECONOMIC STUDIES IN THE GULF OF MEXICO

RICHARD G. HILDRETH

Introduction

This paper reviews the legal framework for Minerals Management Service (MMS) socioeconomic studies carried out as part of the federal offshore oil and gas program in the Gulf of Mexico. Regarding the methodologies to be used for such studies, the relevant legislation and regulations only refer to collecting "time-series and data trend information," designing "experiments to identify" the causes of change, and quantifying impacts (including "cumulative impacts and effects") "to the fullest extent possible including magnitude and duration."

Congressional legislative history documents suggest that production areas be monitored in a manner "designed to provide time series data which can be compared with earlier studies ... for the purpose of identifying any significant changes and the possible cause of such changes" (House Report 95-1590, August 29, 1977). For studies included in environmental impact statements (EISs) prepared by MMS pursuant to the National Environmental Policy Act (NEPA), the applicable regulations state:

> Agencies shall insure the professional integrity, including scientific integrity, of the discussions and analyses in environmental impact statements. They shall identify any methodologies used and shall make explicit reference by footnote to the scientific and other sources relied upon for conclusions in the statement. An agency may place discussion of methodology in an appendix. (40 C.F.R. § 1502.24)

Thus, the choice of methodologies to be used for MMS socioeconomic studies is left largely to the discretion of MMS staff within the very general legal framework discussed next.

Legal Overview

The legal framework for studies in support of the federal offshore oil and gas drilling program has been relatively stable since the major amendments to the 1953 Outer Continental Shelf Lands Act (OCSLA) in 1978 (Hildreth 1976, 1986). It is based on legislative requirements contained in the OCSLA and NEPA and regulations and court decisions interpreting those legislative requirements. Section 20 of the OCSLA requires "a study of any area or region included in any oil and gas lease sale in order to establish information needed for assessment and management of environmental impacts on the human, marine, and coastal environments ..." (43 U.S.C. § 1346(a)(1)). That section also requires additional studies subsequent to the leasing and developing of any area or region to establish environmental information and monitor the human, marine, and coastal environment in a manner designed to provide "time-series and data trend information."

For these purposes the "human environment" is legislatively defined as the "physical, social, and economic components, conditions, and factors which interactively determine the state, condition, and quality of living conditions, employment, and health of those affected, directly or indirectly, by activity occurring on the Outer Continental Shelf" (43 U.S.C. § 1331(i)).

These requirements are implemented through environmental assessment (EA) documents prepared under NEPA Section 102 (42 U.S.C. § 4332) and accompanying Council on Environmental Quality (CEQ) regulations. Section 102(2)(A) requires federal agencies including MMS to "utilize a systematic, interdisciplinary approach which will insure the integrated use of the natural and social sciences" in their decision making which may have an impact on the human environment.

Many NEPA practices and procedures evolved from major federal energy development initiatives onshore and offshore (Harris et al. 2003). EISs prepared as part of the OCS lease sale process in the early 1970s ranged from a 67-page EIS for the leasing of 80 tracts in the Gulf of Mexico to a four-volume, 2000-page EIS for a 1975 Southern California lease sale. Court challenges by environmental and fishing groups and state and local governments to lease sale EISs sometimes resulted in federal court orders delaying the sales pending EIS revisions (Hildreth 1976, 1986). The EIS defects to be remedied usually involved inadequate evaluation of oil spill risks and harms and energy supply and conservation alternatives. In the 1970s and 80s challenges to three Georges Bank lease sales resulted in thirteen separate federal court opinions, many involving NEPA compliance issues. Sale delays usually decreased industry interest in the sale area. They provided sale opponents time to negotiate tract exclusions and sale conditions and seek Congressional sale moratoria. With the most contentious OCS areas off the west and east coasts covered by continuing sale moratoria, NEPA compliance litigation has decreased markedly (Christie and Hildreth 1999).

CEQ recently sponsored a major review of NEPA practices and procedures (Connaughton 2003; Goldfarb 2003). Possible changes discussed to date do not appear to affect MMS assessments of OCS oil and gas development activities.

Legal Requirements for Socioeconomic and Cumulative Impact Assessment

Socioeconomic Effects

The CEQ NEPA regulations state "economic or social effects are not intended by themselves to require preparation of an environmental impact statement. When an environmental impact statement is prepared and economic or social and natural or physical environmental effects are interrelated, then the environmental impact statement will discuss all of these effects on the human environment" (40 C.F.R. § 1508.14). According to the regulations, effects include

70

"aesthetic, historic, cultural, economic, social, or health, whether direct, indirect, or cumulative"; "indirect effects may include growth inducing effects and other effects related to induced changes in the pattern of land use, population density or growth rate, and related effects on air and water and other natural systems, including ecosystems" (40 C.F.R. § 1508.8).

The outer legal limits on socioeconomic assessments are not well defined. In *City of Davis v. Coleman*, 521 F.2d 661 (9th Cir. 1975), the federal appellate court ruled that indirect effects which are merely "speculative" do not have to be evaluated. Otherwise, the federal courts have not provided much guidance on the type and scope of socioeconomic effects which must be evaluated under NEPA. In *State of Louisiana v. Lujan*, 777 F. Supp. 486 (E.D.LA. 1991), the state of Louisiana sued to stop OCS Sale 135 based on the alleged inadequate evaluation of socioeconomic effects in the final NEPA EIS for Sale 135. While there was some evidence that the EIS failed to consider or fully disclose the potential socioeconomic effects of Sale 135, the court found those defects an insufficient basis for issuing an injunction stopping the sale and did not provide any guidelines for improved studies.

CEQ has issued guidance (CEQ 1997) on how federal agencies can comply with Executive Order 12,898 on environmental justice (E.O. 12,898, 59 Fed. Reg. 7629 (1994)). The guidance states that the identification in the NEPA process of disproportionately high and adverse effects (including "multiple or cumulative exposure to human health or environmental hazards") on minority and low-income populations "should heighten agency attention to alternatives (including alternative sites), mitigation strategies, monitoring needs, and preferences expressed by the affected community or population."

Cumulative Impacts

The CEQ NEPA regulations define a cumulative impact as an "impact on the environment which results when the incremental impact of the action when added to other past, present, and reasonably foreseeable future actions regardless of what agency (Federal or non-Federal) or person undertakes such other actions. Cumulative impact can result from individually minor but collectively significant actions taking place over a period of time" (40 C.F.R. § 1508.7). OCSLA regulations state that findings from MMS environmental studies "shall be used to recommend modifications in practices which are employed to mitigate the effects of OCS activities and to enhance the data/information base for predicting impacts which might result from a single lease sale or *cumulative* OCS activities" (30 C.F.R. § 256.82(d)) (emphasis added). Thus OCS lessees submitting exploration and development and production plans for MMS approval are required to provide an assessment of the cumulative effects on the offshore and onshore environments expressed in terms of "magnitude and duration" (30 C.F.R. § 0.203(a)(17); 250.204(a)(11)). These lessee assessments can be helpful to MMS in satisfying its obligations under OCSLA section 20 discussed above.

There have been several judicial interpretations of the cumulative impacts assessment requirement relevant to MMS ESP responsibilities under the OCSLA. In *Natural Resources Defense Council, Inc. v. Hodel*, 865 F.2d 288 (D.C. Cir. 1988), the federal appellate court invalidated Interior Secretary Hodel's approval of the 1987–1992 five-year OCS Lease Sale

Program due to the failure of the final EIS to consider the cumulative impact of simultaneous offshore oil and gas development in the Pacific and Alaska regions on marine species that migrated between those regions, such as whales and salmon. That defect was subsequently corrected, and sales in those regions were carried out pursuant to the program. In *North Slope Borough v. Andrus*, 486 F. Supp. 332 (D.D.C. 1980), rev'd on other grounds, 642 F.2d 589 (D.C. Cir. 1980), involving Alaska OCS oil and gas leasing, both the trial and appellate courts ruled that the cumulative effects of "other significant Federal and state energy development projects ... in progress and planned for the North Slope Region" had to be considered in the EIS.

A widely-quoted decision of the federal appellate court for the Central and Western Gulf of Mexico region has stated that an EIS cumulative-effects study must identify: 1) the area in which effects of the proposed project will be felt; 2) the impacts that are expected in that area from the proposed project; 3) other actions—past, proposed, and reasonably foreseeable—that have had or are expected to have impacts in the same area; 4) the impacts or expected impacts from these other actions; and 5) the overall impact that can be expected if the individual impacts are allowed to accumulate (Fritiofson v. Alexander, 772 F.2d 1225 (5th Cir. 1985)). In addition, every three fiscal years, OCSLA section 20(e) requires the Interior Secretary to submit to Congress and the public "an assessment of the cumulative effect" of OCS oil and gas activities on the "human, marine, and coastal environments" (43 U.S.C. § 1346(e)).

All studies of cumulative environmental impacts should include associated socioeconomic effects (Bowen and Riley 2003; Kildow et al. 2000).

MMS Implementation

MMS's OCS Environmental Studies Program's (ESP's) principal function is to obtain information about the ecological, physical, oceanographic, and socioeconomic implications of energy resource development (Murdock et al. 2002). It supports MMS and Interior Department decision making at all stages of the offshore oil and gas leasing program carried out under the OCSLA in compliance with the foregoing impact assessment requirements including the post lease sale stages of exploration (30 C.F.R. § 250.203(g)); development and production (30 C.F.R. § 250.204(j)); and decommissioning (30 C.F.R. § 250.1700). However, for mature areas of the OCS such as the Central and Western Gulf of Mexico, MMS has determined that environmental assessments generally are not required for the post lease sale stages of exploration plans and development and production plans, 51 F.R. 1855, 1857, Jan. 15, 1986, and lessee information gathering obligations are correspondingly reduced (30 C.F.R. § 0.203(a)(8)). A CEQ NEPA regulation (40 C.F.R. § 1508.4) authorizes such "categorical exclusions" from the NEPA process and agency applications of the regulation are judicially reviewable. California v. Norton, 311 F.3d 1162 (9th Cir. 2002). Exclusion from NEPA of development and production plans reviewed by MMS in the Gulf is to some extent supported by OCSLA sections 25(e)(1) and (l) (43 U.S.C. § 1351(e)(1), (l)).

On June 27, 2002, Interior Secretary Norton approved the OCS Oil and Gas Leasing Program for 2002–2007 (MMS 2002b). The two-volume final EIS (MMS 2002a) accompanying the program contains environmental, socioeconomic, and cumulative impact information for comparing the three regions covered by the program, Pacific, Alaska, and the Gulf of Mexico. The program includes annual area-wide sales in the Central and Western Gulf of Mexico and two sales in the Eastern Gulf. Assessing the environmental and associated socioeconomic and cumulative impacts of these sales is the responsibility of MMS's Gulf of Mexico Regional Office. The final EISs for Lease Sale 181 in the Eastern Gulf (MMS 2001a) and for nine 2003–2007 Central and Western Gulf lease sales (MMS 2002c) contain extensive socioeconomic and cumulative impact analyses as well as short discussions of environmental justice issues. The brief socioeconomic and cumulative impact analyses in the EA documents for Western Gulf sales 180 and 184 (MMS 2001b, 2002d) build on the more detailed analyses contained in the May 1998 final EIS for Proposed Lease Sales 171, 174, 177, and 180. Recent community (Hughes et al. 2002; Keithly 2001) and regional (Wallace et al. 2001) level background studies commissioned by MMS's Gulf regional office have focused on socioeconomic rather than cumulative impacts.

In addition to the cumulative impacts they already discuss, Gulf of Mexico Lease Sale EISs could evaluate the cumulative impacts of activities such as offshore oil exploration, development, production, and facilities decommissioning (Lima 1997) activities in Mexican waters of the Gulf, and possible new oil and gas transport facilities on the Gulf OCS to be licensed by the Coast Guard under the Deep Water Port Act as amended in 2002 (Public Law 107–295, November 25, 2002)). To include Mexico's Gulf waters in MMS cumulative impact assessments is supported by a 1997 CEQ memorandum (McGinty 1997) and several lower federal court decisions (Goldfarb 2003).

In addition, platform removals as part of offshore oil decommissioning activities in state and federal waters off California have raised concerns about the spread of marine invasive species, an issue also of concern in the Gulf (Showalter 2003).

On the other hand, mining of methane hydrate deposits on the Gulf sea floor is probably too "speculative" to require any sort of detailed analysis of potential cumulative impacts at this time. Less speculative would be evaluation of the cumulative impacts on Gulf of Mexico ocean resource users including offshore oil operators of existing and proposed marine protected areas and marine reserves (Carter 2003).

Conclusion

Past socioeconomic and cumulative impact assessments carried out by MMS in the Gulf have fit relatively comfortably within the applicable legal framework. Changes in methodology and emphasis in future assessments currently being discussed by MMS staff (Cluck and Luton 2002a, 2002b; Luton and Cluck 2003; Murdock et al. 2002) could improve the social scientific validity and utility of the assessments while continuing to meet MMS legal obligations. To the extent the proposed changes facilitate increased tracking of cumulative impacts, they will assist MMS in meeting key legal obligations under the National Environmental Policy Act and the Outer Continental Shelf Lands Act.

References

Bowen, R.E. and C. Riley. 2003. Socio-economic Indicators and Integrated Coastal Management. Ocean & Coastal Management 46:299–312.

Carter, D.W. 2003. Protected Areas in Marine Resource Management: Another Look at the Economics and Research Issues. Ocean & Coastal Management 46:439–456.

Christie, D.R. and R.G. Hildreth. 1999. Coastal and Ocean Management Law in a Nutshell. St. Paul, MN: West Group.

Cluck, R.E. and H. Luton. 2002a. Responding to Social Change and Avoiding Self-Subversion: The Social and Economic Effects of Offshore Oil and Gas Development on Communities, Families, and Individuals in the Gulf of Mexico. *In* SPE International Conference on Health, Safety & Environment in Oil and Gas Exploration and Production Proceedings: 20–22 March, 2002, Kuala Lumpur, Malaysia. CD-ROM. Dallas, TX: Society of Petroleum Engineers.

Cluck, R.E. and H. Luton. 2002b. Towards a Multi-Level Social Assessment Framework: Effects and Responses to Change in the Gulf of Mexico. *In* SPE International Conference on Health, Safety & Environment in Oil and Gas Exploration and Production Proceedings: 20–22 March, 2002, Kuala Lumpur, Malaysia. CD-ROM. Dallas, TX: Society of Petroleum Engineers.

Connaughton, J.L. 2003. A More Effective and Timely NEPA. *In* Proceedings of the Rocky Mountain Mineral Law Forty-Ninth Annual Institute, Chap. 2. Westminster, CO: Rocky Mountain Mineral Law Foundation.

Council on Environmental Quality (CEQ). 1997. Environmental Justice: Guidance under the National Environmental Policy Act. Washington, DC: CEQ. http://www.epa.gov/compliance/resources/policies/ej/ej_guidance_nepa_ceq1297.pdf (accessed December 18, 2003).

Goldfarb, D. 2003. NEPA: Application in the Territorial Seas, the Exclusive Economic Zone, the Global Commons, and Beyond. Southwestern University Law Review 32:735–760.

Harris, C., R. Tuchman, and J. Hall. 2003. New Brine, Old Bottle: NEPA and NPDES Bottlenecks and Potential Solutions in Coalbed Methane Production in the Powder River Basin. *In* Proceedings of the Rocky Mountain Mineral Law Forty-Ninth Annual Institute, Chap. 13. Westminster, CO: Rocky Mountain Mineral Law Foundation.

Hildreth, R.G. 1976. The Coast: Where Energy Meets the Environment. San Diego Law Review 13:253–305.

Hildreth, R.G. 1986. Ocean Resources and Intergovernmental Relations in the 1980s: Outer Continental Shelf Hydrocarbons and Minerals. *In* Ocean Resources and U.S. Intergovernmental Relations in the 1980s. Maynard Silva, ed. Boulder, CO: Westview Press, pp. 155–196.

Hughes, D.W., J.M. Fannin, W. Keithly, W. Olatubi, and J. Guo. 2002. Lafourche Parish and Port Fourchon, Louisiana: Effects of the Outer Continental Shelf Petroleum Industry on the Economy and Public Services, Part 2. New Orleans, LA: U.S. Department of the Interior, Minerals Management Service, Gulf of Mexico OCS Region. OCS Study MMS 2001-020.

Keithly, D.C. 2001. Lafourche Parish and Port Fourchon, Louisiana: Effects of the Outer Continental Shelf Petroleum Industry on the Economy and Public Services, Part 1. New Orleans, LA: U.S. Department of the Interior, Minerals Management Service, Gulf of Mexico OCS Region. OCS Study MMS 2001-019.

Kildow, J.T., B. Baird, C.S. Colgan, H. Kite-Powell, and R. Weiher. 2000. Improving Economic Information for Ocean and Coastal Resources. *In* Coasts at the Millennium: Proceedings of the Seventeenth International Conference of the Coastal Society, 9–12 July 2000, the Portland Marriott Riverfront, Portland, Oregon, L.W. Jodice, A. Gupta, and R. Boyles, eds. Alexandria, VA: The Coastal Society, pp. 305–313.

Lima, J. 1997. Long-term Socioeconomic Effects Of Onshore Facility Decommissioning. *In* Decommissioning and Removal of Oil and Gas Facilities Offshore California: Recent Experiences and Future Deepwater Challenges. F. Manago and B. Williamson, eds. Santa Barbara, CA: University of California, Santa Barbara, Marine Science Institute, pp. 124–135.

Luton, H. and R.E. Cluck. 2003. Social Impact Assessment and Offshore Oil and Gas in the Gulf of Mexico. Draft. N.p.: U.S. Department of the Interior, Minerals Management Service. http://www.mms.gov/eppd/socecon/files/gulfSIA.pdf (accessed December 8, 2003).

McGinty, K.A. 1997. Memorandum to Heads of Agencies on the Application of the National Environmental Policy Act to Proposed Federal Actions in the United States with Transboundary Effects. Council on Environmental Quality, Washington, DC, July 1. http://ceq.eh.doe.gov/nepa/regs/transguide.html (accessed December 18, 2003).

Minerals Management Service (MMS). 2002a. Outer Continental Shelf Oil & Gas Leasing Program: 2002–2007: Final Environmental Impact Statement. 2 vols. Herndon, VA: U.S.

Department of the Interior, Minerals Management Service. OCS EIS/EA MMS 2002-006.

Minerals Management Service (MMS). 2002b. Proposed Final Outer Continental Shelf Leasing Program 2002–2007. Herndon, VA: U.S. Department of the Interior, Minerals Management Service.

Minerals Management Service (MMS). Gulf of Mexico OCS Region. 2002c. Gulf of Mexico OCS Oil and Gas Lease Sales, 2003-2007: Central Planning Area Sales 185, 190, 194, 198, and 201: Western Planning Area Sales 187, 192, 196, and 200: Final Environmental Impact Statement. 2 vols. New Orleans, LA: U.S. Department of the Interior, Minerals Management Service, Gulf of Mexico OCS Region. OCS EIS/EA MMS 2002-052.

Minerals Management Service (MMS). Gulf of Mexico OCS Region. 2002d. Proposed OCS Lease Sale 184, Western Gulf of Mexico: Environmental Assessment. New Orleans, LA: U.S. Department of the Interior, Minerals Management Service, Gulf of Mexico OCS Region. OCS EIS/EA MMS 2002-008.

Minerals Management Service (MMS). Gulf of Mexico OCS Region. 2001a. Gulf of Mexico OCS Oil and Gas Lease Sale 181, Eastern Planning Area: Final Environmental Impact Statement. 2 vols. New Orleans, LA: U.S. Department of the Interior, Minerals Management Service, Gulf of Mexico OCS Region. OCS EIS/EA MMS 2001-051.

Minerals Management Service (MMS). Gulf of Mexico OCS Region. 2001b. Proposed OCS Lease Sale 180, Western Gulf of Mexico: Environmental assessment. New Orleans, LA: U.S. Department of the Interior, Minerals Management Service, Gulf of Mexico OCS Region. OCS EIS/EA MMS 2001-034.

Murdock, S.H., F.L. Leistritz, and S. Albrecht. 2002. An Examination of Selected Recent MMS Socioeconomic Studies and Assessments in the Gulf of Mexico. Herndon, VA: U.S. Department of the Interior, Minerals Management Service, Environmental Studies Program. http://www.mms.gov/itd/pubs/2002/2002-057.pdf (accessed January 20, 2004).

Showalter, S. 2003. Aquatic nuisance species in the Gulf of Mexico: A Guide for Future Action by the Gulf of Mexico Regional Panel and the Gulf States. University, MS: Sea Grant Law Center, University of Mississippi. http://www.olemiss.edu/orgs/SGLC/ANS.pdf (accessed December 5, 2003).

Wallace, B., J. Kirkley, T. McGuire, D. Austin, and D. Goldfield. 2001. Assessment of Historical, Social, and Economic Impacts of OCS Development on Gulf Coast Communities. 2 vols. New Orleans, LA: U.S. Department of the Interior, Minerals Management Service, Gulf of Mexico OCS Region. OCS Study MMS 2001-026 and 2001-027.

APPENDIX D2

THE HISTORY OF THE OFFSHORE INDUSTRY IN THE GULF OF MEXICO

TYLER PRIEST

In their "Social Impact Assessment" essay, Harry Luton and Rodney Cluck have a section entitled: "Challenge of Identifying the Offshore Oil Industry." This challenge, they write, is due to "the size and complexity of this industry, because the full spectrum of enterprises involved in finding, extracting, processing, storing, and bringing petroleum-based products to market is present, because the support and transportation requirements of offshore operations add substantially to the complexities and variabilities of the oil industry, and because of its uneven distribution across the Gulf" (Luton and Cluck 2004).

Indeed, the offshore industry in the Gulf has a long, varied, and dynamic history. When scholars first attempted in the mid-1980s, during the oil bust, to make some sense of this history, they assumed it was over, or at least winding down. The Gulf was written off by many people as the "Dead Sea." Then came the deepwater miracle, which initiated yet another era in this history and forced us to reevaluate early conclusions and rethink how we apply received wisdom about extractive economies, at least in the case of the Gulf. Luton and Cluck ask some excellent questions about how we identify the industry, historically and otherwise. Offshore development in the Gulf has been unique, but it also has served as a model for other offshore provinces. Unlike most petroleum provinces, in which discoveries have been concentrated in a short span of one to three decades, substantial discoveries have been made in the Gulf of Mexico basin for a century. In contrast to the major provinces of the world, where hydrocarbons are concentrated in a small number of world-class "giant" fields (fields with a known recovery of 500 million barrels of oil equivalent [boe] or more), the GOM basin has yielded thousands of smaller fields as well as numerous giants and "large" fields of 50 to 500 million boe (Nehring 1991), which has made room for both large and small operators and a brilliant diversity of contractors. As the search for oil and gas around the world moved into deeper waters and new offshore environments over the last thirty years, the Gulf of Mexico has remained the primary laboratory for technological innovation and regulatory practices.

Concerned with the social and economic impacts of offshore development, scholars and analysts of the industry have rightfully focused on local and micro-level changes affecting communities, support-centers, fabrication yards, demographics of the labor force, etc. What Luton and Cluck appear to be calling for in their paper, and what we have been trying to do in the GOM history project, is marry our knowledge of the local-level effects with a macro-level understanding of changes in the oil industry and, I would add, in business-government relations. So I will try to offer some provisional and very general thoughts on this.

It is important to keep in mind two things when looking at this history. First, the ongoing search for oil in the GOM has been a function of the long-term decline of the lower 48 United States as a petroleum province. The discovery of oil in this country peaked in 1930; oil production peaked in 1970. By the end of World War II, the most obvious oil onshore, contained in structural traps, had already been found. To replace depleted reserves, new kinds of traps and oil would have to be found using new exploration concepts and technologies and in higher cost environments. A good way of characterizing the post-World War II period, in my view, is as a "race against depletion." Offshore GOM is the feature development in all of this. The second point, related to the first, is that the maturation of offshore oil and gas has occurred during a period, especially since 1973, of incredible volatility for the petroleum industry, involving soaring and plunging prices linked to the influx of foreign oil into the United States, new waves of government regulation, and drastic corporate restructuring. Only continuous revolutions in technology have sustained the GOM through this turbulent period.

First Era: The New Frontier, 1938–1962

In the late 1930s, major oil companies, through the corrupt leasing practices in Louisiana, obtained hugely profitable oil and gas fields in south Louisiana. This combined with the doubling of base domestic oil prices upon the lifting of price controls at the end of WWII both generated interest in the adjacent offshore domain and financed a new wave of exploration and drilling by the established firms (Humble [Exxon], Gulf, Shell, and California [Chevron] being the most active). Meanwhile, other oil firms who had missed out on the big action in south Louisiana (Kerr-McGee, Pure, Magnolia, Union, Superior) also cast their sights into shallow open water. The opportunism of oilmen was rewarded. During 1949–1956, the increase to U.S. domestic reserves from offshore development was nine times the average for onshore wells (U.S. Department of the Interior 1969).

Technological innovations in geophysical exploration, drilling, and marine design and construction held the keys to this success. Operating in an open-water marine environment entailed great risk and required specialized expertise. To meet the challenge, the oil companies sponsored research at the major universities in this region and hired a generation of scientists and engineers out of them. And they both tapped into a preexisting Gulf Coast oil-service sector and cultivated competitively organized new ones in drilling and supply, transferring some of this risk to others and protecting themselves against the high infrastructure costs and irregular pace of exploration and development (Baxter 1997). The seeds of the larger offshore industry were planted during this period with the emergence of geophysical contractors (i.e. GSI, Western, Petty-Ray); engineering and construction firms (Brown & Root, J. Ray McDermott); supply and transport firms (Tidewater, Petroleum Helicopters); naval architects (Friede-Goldman); and various shipyards to construct new-fangled drilling vessels. The drilling companies, such as ODECO, Zapata, Global Marine, The Offshore Company, captured imaginations with the innovative development of a variety of submersible and jack-up rigs.

The industry loves to celebrate the entrepreneurship of these early pioneers, and rightly so. But I would like to point out that government aid and support for the industry were also crucial to offsetting high costs and risks. These include not only the well-known tax benefits and incentives the industry as a whole received in the United States—such as the percentage depletion allowance and import quotas, imposed on a mandatory basis in 1959, just as major offshore discoveries were coming on stream—but specific assistance in the form of government technology transfers (i.e. sonar and radio-positioning; diving; war-surplus vessels; generous prorationing; little or no safety or environmental regulations). And despite oil company fears that the triumph of the federal government in the legal showdown with Louisiana and other states over control of the "tidelands" in the 1950s would harm prospects for development, federal leasing, begun in 1954, introduced an orderly method for cheaply and efficiently transferring huge tracts of the public offshore domain to industry.

Second Era: Emergence of an Industry, 1962–1973

Despite federal support, in the late 1950s, many people, even those intimately involved with offshore, saw the GOM as a unique and perhaps temporary opportunity, not something that had long-term viability. In the late 1950s, Bouwe Dykstra, Shell Oil's New Orleans vice president who was almost single-handedly responsible for carving out Shell's strong position in the shallow-water Gulf, adamantly insisted that development would never reach beyond 200-foot depths. Not until the 1960s did the various companies involved offshore envision a longer future and come together as an industry, bound together by a common technological purpose.

Two events in the early 1960s helped foster this cohesion. First was the historic March 1962 offshore lease sale, in which the federal government offered for lease every tract nominated, awarding more than 2 million acres in leases and collecting $445 million in bonus bids, more than all the timber sales in Oregon and California and onshore mineral leasing for the year combined (Rankin 1986). Companies immediately began scrambling to drill their large inventory of leases, setting off a major boom in all related service and support businesses. The 1962 sale not only provided more leases for a larger number of companies to choose from, but it also drove down the price of cash bonuses, allowing smaller independents, who had begun to be shut out of the offshore game, to acquire a piece of the action, albeit still dominated by the majors (Lohrenz and Oden 1973). The sale also helped jump start major exploration for natural gas in the Gulf.

The second major event was Shell Oil's launching of the first semi-submersible drilling vessel, the *Bluewater 1*, and installation of the first subsea well, which greatly extending exploration into deeper waters. In 1963, to help spread offshore capabilities, Shell Oil decided to share these revolutionary technologies with the industry. Shell's "million dollar school" signaled the beginning of greater cooperation and consultation within the industry to meet difficult technological challenges (Taylor 1983). Three major hurricanes in the 1960s (Hilda, Betsy, and Camille) also helped bring about a convergence of improved ideas and practices on platform design and construction, allowing production technology to begin catching up with exploration technology, culminating with the founding of the Offshore Technology Conference in 1969. Greater industry consultation on technical matters also helped companies produce the first "recommended practices" (RP) documents and meet rising health and safety and environmental regulations in the 1970s.

Despite great progress, by the end of the 1960s the industry faced limits to its expansion. Although enough important discoveries were made to hold oil company interest, many leases proved to be unproductive and the cost of bringing in the productive ones began to outrun the price of oil, which had stubbornly remained in the $2–3 barrel range. Furthermore, the federal government moved to ration leases more severely under the so-called "tract selection" system in search of greater revenues with which to help pay for the Vietnam War. Tract selection offered blocks in a piecemeal fashion. Given escalating costs in deeper water, this hindered more efficient exploration strategies involving basin-wide geological assessments or structural trends or plays that transcended tract boundaries, thus constraining profitable development. A 1978 study showed that since the beginning of federal leasing through 1975, the industry had paid out $18 billion for bonuses but had only earned $17.8 billion in revenues. Many leases for which bonus figures were included in this statistic had yet to pay out, but the point remained that success in the Gulf of Mexico had been highly variable. Notable successes included Shell Oil, Gulf Oil, Chevron, Tenneco, and Forest Oil, while striking failures were Texaco, Amoco, and Sun Oil (Dougherty et al. 1978).

Third Era: Feeding Frenzy, 1973–1985

In the early 1970s, the major contradiction of the early offshore system in the Gulf was temporarily resolved. Most obviously, the price spike caused by the OPEC oil embargo in 1973 made offshore a much more profitable endeavor. Companies could afford a higher ratio of dry holes and unproductive leases to discoveries, even in progressively deeper water. However, lease bonuses began rising sharply before 1973, so this trend cannot be solely attributed to rising oil prices, improved production technology, or to the federal policy of tract selection. An unappreciated factor in this trend was the discovery and adoption of the "bright spot" or "direct detection" method of interpreting seismic data. Advanced digital recording and processing of seismic data, which had made quantum leaps in the mid-1960s, enabled geophysicists to detect hydrocarbons on the seismic record. Bright spots greatly diminished the dry hole factor in the risk equation, allowing companies to put a lot more money into its lease bids and more than make up for it in decreased drilling costs. Once the technology was developed and embraced, it had a giant impact on offshore exploration in the Gulf and helped pioneer a new deeper water hydrocarbon trend and paved the way for the installation of a new wave of massive fixed platforms (Forrest 2000).

By the early 1980s, however, this adaptation appeared to give the industry only a short lease on life in the Gulf. The price of lease bonuses had risen to astronomical levels. In OCS Sale A62 in 1980, Superior-Pennzoil-Sohio paid $165 million for a single tract, Ewing Bank 304. This sum dwarfed Shell Oil's winning $10,000 bid in 1947 for South Pass Block 24, the largest offshore field discovered in the Gulf for many years ($52,457/acre versus $4/acre). Even the largest firms could not afford to bid alone, and had to bring in partners to offload some of the capital risk—and, although they would not admit this, reduce the competition. Despite the price spike of the "second oil shock" and the application of new technologies, offshore leases were becoming prohibitively expensive. So was the cost of development as the deepwater frontier migrated out to the edge of the continental shelf in 1,000-feet-plus depths. In 1981, oil prices peaked and began to plunge. Industry leaders complained that the tract selection system of leasing was creating a shortage of exploration opportunities in the declining oil province of the United States, especially as environmental concerns blocked leasing off the Atlantic and Pacific coasts (Farrow 1990: 137–38).

Fourth Era: Deepwater, 1985–Present

The collapse in oil prices in the mid-1980s decimated the offshore industry in the Gulf, especially the drilling, service, and supply companies who assumed a great deal of the infrastructure and economic risk of offshore operations. In the midst of the carnage, the introduction in 1983 of a new "area-wide leasing" (AWL) system was the first step in reviving the "Dead Sea." AWL put into play entire planning areas (e.g., the central Gulf of Mexico) up to 50 million acres, as opposed to tracts specifically nominated and offered under the tract selection system. Some critics have charged that AWL stifled competition and that it has received too much credit for the deepwater boom of recent years. Proper credit, they insist, goes to technological innovations such as 3-D seismic, tension-leg platforms, directional drilling, and subsea wells (Boué and Luyando 2002). Yet, not until companies possessed cheap and extensive acreage did they have the incentive to develop and refine these technologies.

Competition was indeed muted early in the game. Deepwater is too expensive, risky, and dangerous for small or undercapitalized companies. Shell Oil acquired a vast majority of the early deepwater leases under AWL and pioneered the technologies needed to operate on them. Recently, much credit for the deepwater boom has been given to the Royalty Relief initiative passed in 1995, but the real stimulus to deepwater exploration was when Shell brought in wells producing 10,000 barrels/day at its Auger prospect in 1994, compared to 500–1,000 b/d for a good well on the shelf. This completely changed the cost structure of the deepwater play. An operator could drastically reduce the number of expensive wells on a given platform and still produce at a greater rate than original estimates required for making the field profitable (Oil & Gas Journal 1995).

The news from Auger set off a flurry of leasing in deepwater by the larger companies. With big discoveries in deepwater, they began to "high-grade" their offshore portfolio, selling off older, producing leases on the shelf to smaller companies with lower overhead and new technological capabilities (examples include Devon, Newfield, Anadarko, Forest) who can still produce those fields profitably. Shell and other super majors such as British Petroleum have established themselves as "basin masters" in deepwater through their control of platform and pipeline infrastructure, but the incredible profitability of deepwater oil and the spread of deepwater engineering technology (floating production systems, subsea wells) have drawn numerous companies, large and small, into the game. Kerr-McGee is an example of a smaller independent who has taken the lead in deepwater. As of 2002, at least forty different operators had drilled deepwater wells in the Gulf (Godec et al. 2002).

81

At the end of each era described above, the offshore industry approached material and economic limits, which were overcome by technological innovation and/or improved terms of access, or in the case of the deepwater era, the discovery of new, highly productive reservoirs. Innovation has come from a wide spectrum of sources, including lessons learned overseas in places like the North Sea, Brazil, and West Africa, the result of this very fluid and dynamic industry bound by a sense of common purpose. The federal government, as well, deserves credit for managing the trade-offs over leasing submerged lands by giving the industry enough access and incentives to make offshore development viable while attempting in with varying degrees of success to maximize revenues and protect the interests of many different groups. I hesitate to venture an analysis of the current situation or make predictions about the future other than to say that of the relatively limited investment opportunities facing private oil companies in the world today—including oil and gas in the Caspian and Russia, LNG, heavy oil, piped natural gas, and the redevelopment of conventional oil in OPEC countries—deepwater oil in the GOM and elsewhere, such as West Africa, is the most profitable of all. Although the GOM cannot reverse the long-term decline of production in the United States, it has helped prolong it and has given private oil companies leverage in the world oil market and in their relationship with OPEC.

References

Baxter, V. 1997. The Effects of Industry Governance on Offshore Oil Development in the Gulf of Mexico. International Journal of Urban and Regional Research, 21(2,1997): 238–258.

Boué, J.C. and G. Luyando. 2002. U.S. Gulf Offshore: Petroleum Leasing and Taxation and Their Impact on Industry Structure, Competition, Production, and Fiscal Revenues. Oxford: Oxford Institute for Energy Studies.

Dougherty, E.L., L.A. Bruckner, and J. Lohrenz. 1978. Cumulative Bonus and Production Profiles with Time for Different Competitive Bidders: Federal Offshore Oil and Gas Leases. American Institute of Mining, Metallurgical and Petroleum Engineers, Society of Petroleum Engineers (SPE) Preprint 7134.

Farrow, S. 1990. Managing the Outer Continental Shelf Lands. New York: Taylor & Francis.

Forrest, M. 2000. Bright Idea Still Needed Persistence. AAPG Explorer (May 2000), AAPG website, http://www.aapg.org/explorer.

Godec, M.L., V.A. Kuuskraa, and B.T. Kuck. 2002. How U.S. Gulf of Mexico Development, Finding, Cost Trends Have Evolved. Oil & Gas Journal (May 6, 2002): 52–60.

Lohrenz, J. and H.A. Oden. 1973. Bidding and Production Relationships for Federal OCS Leases: Statistical Studies of Wildcat Leases, Gulf of Mexico, 1962, and Prior Sales. American Institute of Mining, Metallurgical and Petroleum Engineers, Society of Petroleum Engineers (SPE) Preprint 4498.

Luton, H. and R. Cluck. 2004. Social Impact Assessment and Offshore Oil and Gas in the Gulf of Mexico. Proceedings of the 24th Annual Conference of the International Association of Impact Assessment, 26–29 April 2004, Vancouver, Canada.

Nehring, R. 1991. Oil and Gas Resources. *In* The Gulf of Mexico Basin: The Geology of North America, Volume J.A. Salvador ed. Boulder, CO: The Geological Society of America, 1988.

Rankin, J. 1986. Untitled Manuscript on the History of Federal Offshore Leasing.

Oil & Gas Journal. 1995. Shell Marks Progress in Deepwater Gulf. Oil & Gas Journal On-line (November 13, 1995), http://ogj.pennnet.com.

Taylor, D.M. 1983. 20th Anniversary of Shell's Million-Dollar School of Offshore Technology, Ocean Industry (April 1983): 35–38.

U.S. Department of the Interior. 1969. Petroleum and Sulfur on the U.S. Continental Shelf, August 1969, Box 134, Central Files, 1969–1972, Records of the Secretary of the Interior, Record Group 48, National Archives and Records Administration (NARA), College Park, MD.

APPENDIX D3

PLANNING CHALLENGES AT THE BUREAU OF LAND MANAGEMENT

ROB WINTHROP

Slide 1

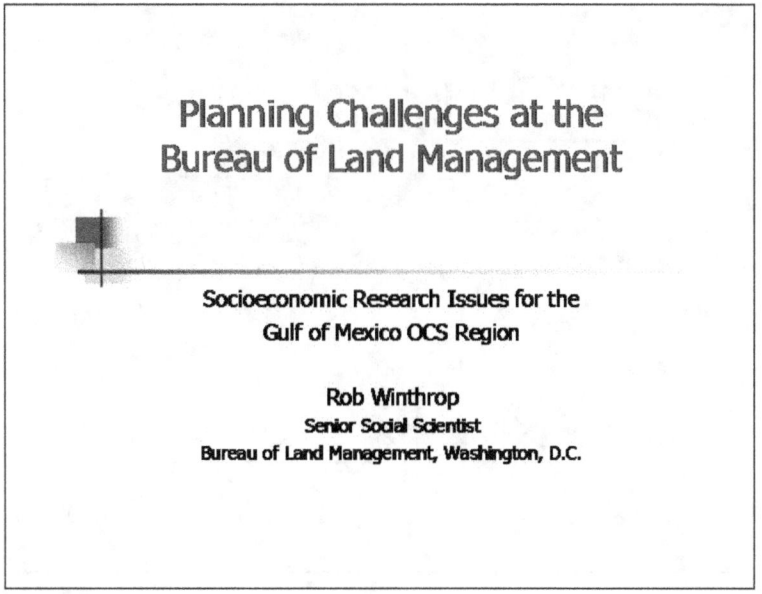

Slide 2

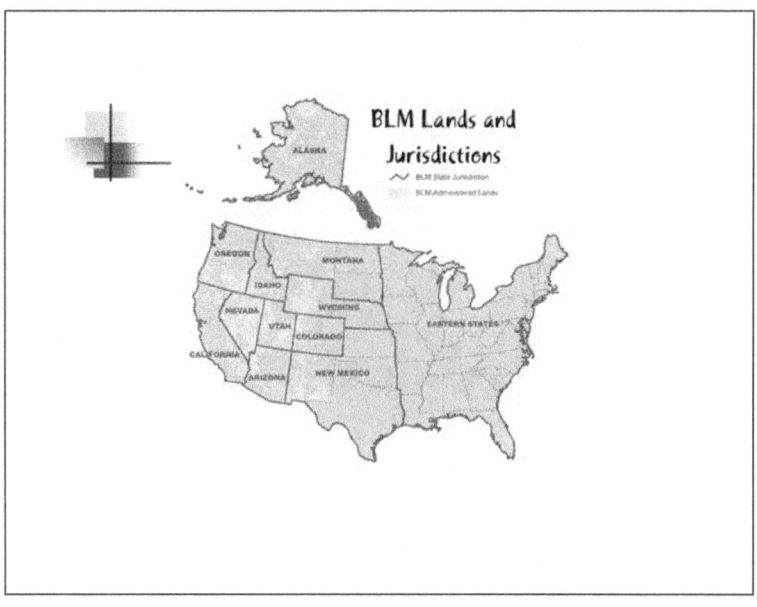

Slide 3

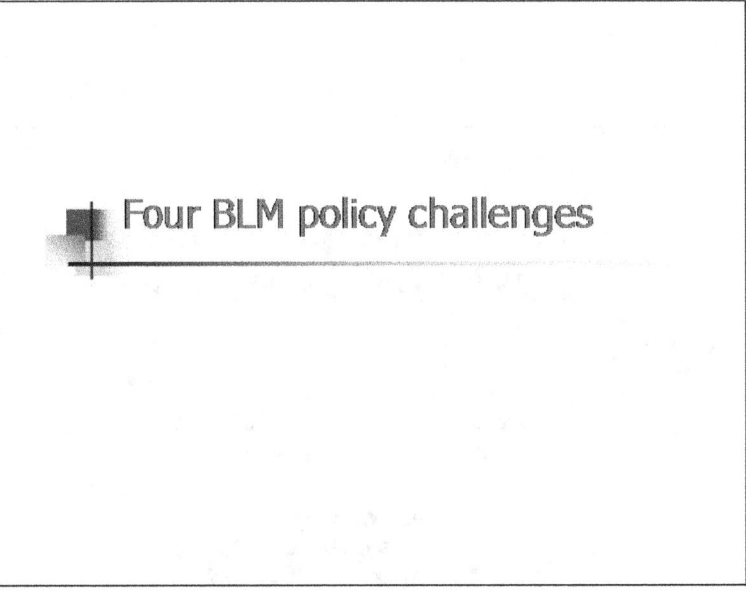

Slide 4

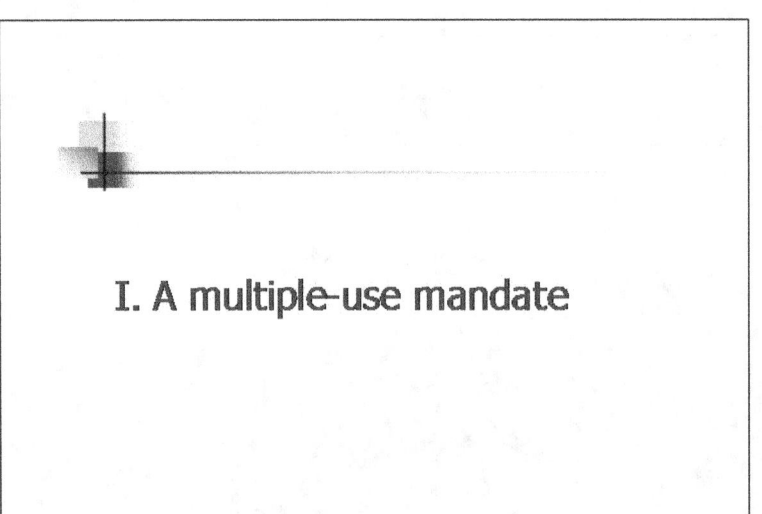

Slide 5

Energy & Minerals

- BLM administers some 700 m acres of federal on-shore mineral estate
- 2.1 trillion cubic feet of natural gas came from onshore public lands in FY 2002
- Over 100 million barrels of oil came from onshore public lands in FY 2002

Slide 6

Renewable Resources

- Rangelands
  - Some 18,000 grazing permits and leases
  - 12.7 million animal-unit months annually
- Timber
  - 32.5 m. cubic feet of timber volume

Slide 7

Conservation

Multiple conservation responsibilities:
- Plants
- Wildlife
- Fisheries
- Rangelands
- Cultural and historic resources
- Paleontological resources

Slide 8

Recreation

- 67 million visitor days (FY02)
- 13 National Conservation Areas
  - California Desert NCA (9.5 m acres)
- 15 National Monuments
  - Grand Staircase Escalante Nat'l Monument (Utah)
- 148 wilderness areas (6.3 m acres)

Slide 9

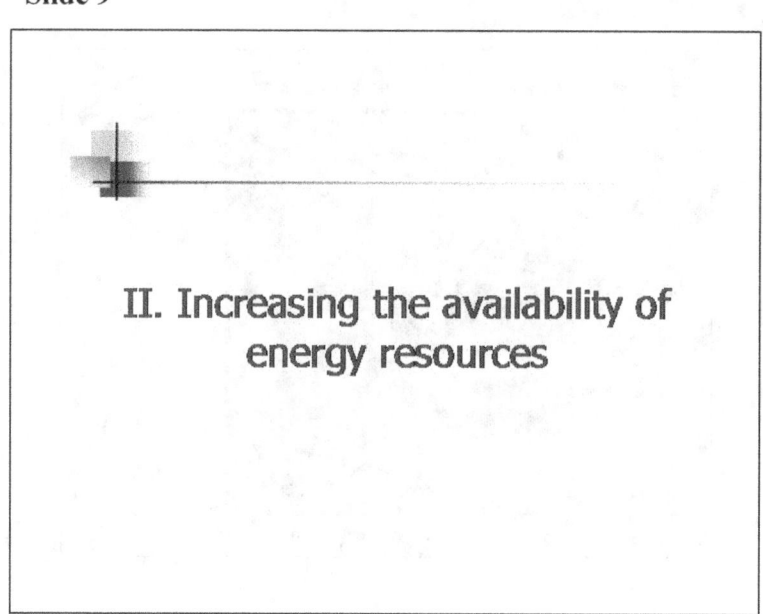

II. Increasing the availability of energy resources

Slide 10

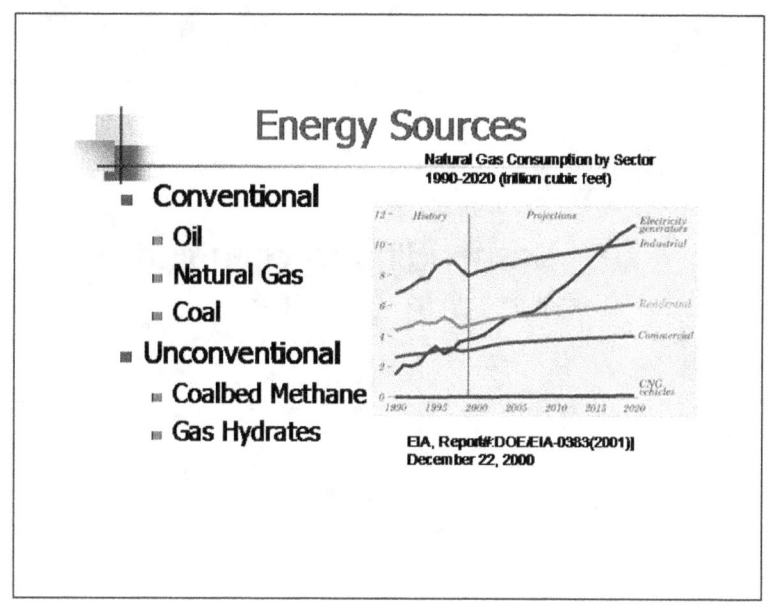

Energy Sources

- Conventional
  - Oil
  - Natural Gas
  - Coal
- Unconventional
  - Coalbed Methane
  - Gas Hydrates

Natural Gas Consumption by Sector
1990-2020 (trillion cubic feet)

EIA, Report#:DOE/EIA-0383(2001)|
December 22, 2000

Slide 11

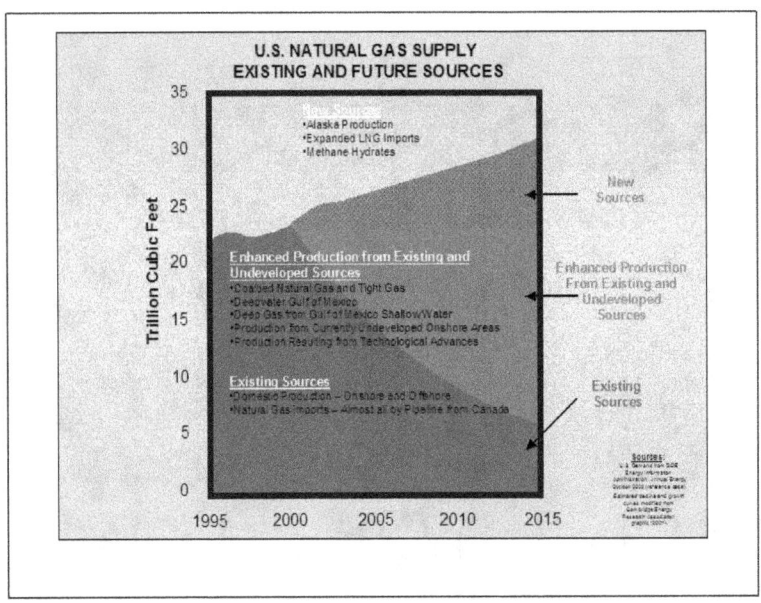

Slide 12

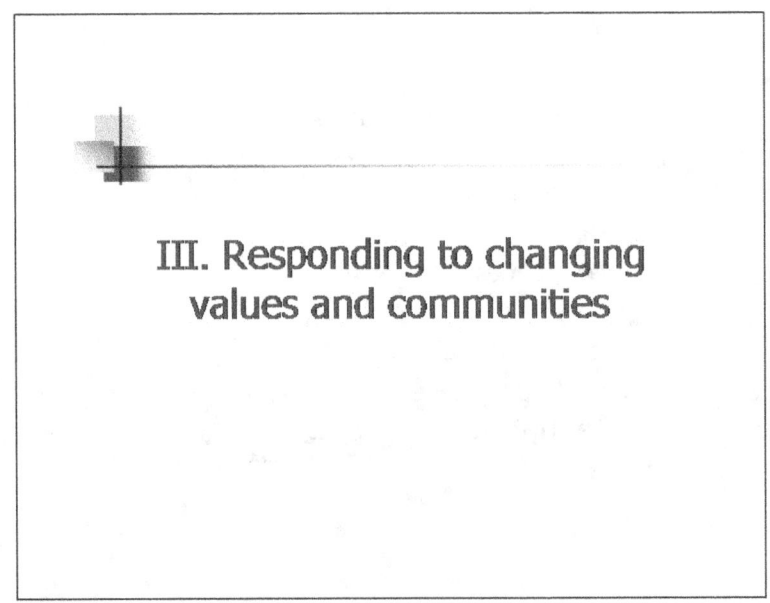

Slide 13

- Growth of environmental values
- The 'changing West' – transformation of western communities and economies
- Expansion of western cities makes BLM lands the backdrop for urban development (Phoenix, Las Vegas)

Slide 14

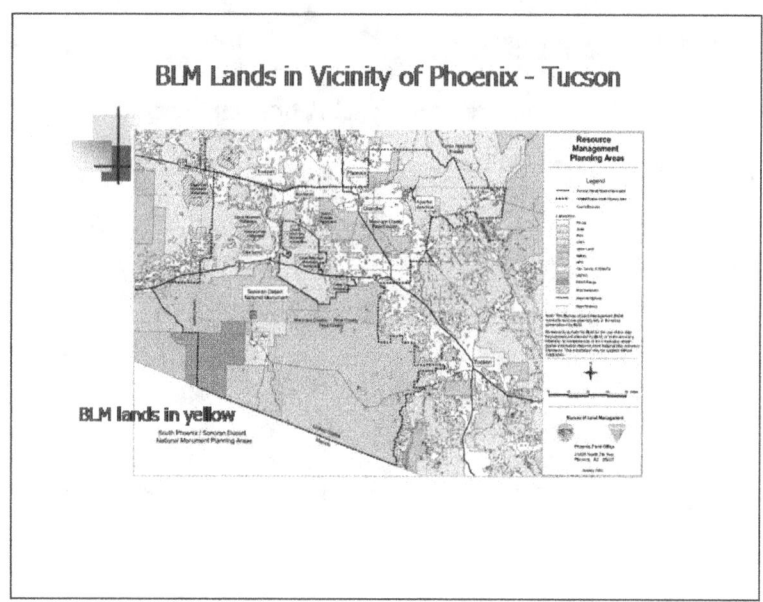

BLM Lands in Vicinity of Phoenix - Tucson

BLM lands in yellow

Slide 15

IV. Emphasis on collaborative
approaches

Slide 16

Scoping meeting: National Petroleum Reserve —
Alaska

Slide 17

- Collaboration with cooperating agencies (local, state, tribal governments) in resource management planning
- Use of partnerships
  - Role of Sonoita Parternship in establishing the Las Cienegas Nat'l Conservation Area

Slide 18

- Increased reliance on alternative dispute resolution policies and procedures
- Use of Economic Profile System as a tool to foster discussion with the public over economic trends and strategies

Slide 19

The challenge for BLM's social impact assessment is not only to assess proposed uses against baseline conditions, but against potential competing uses.

Slide 20

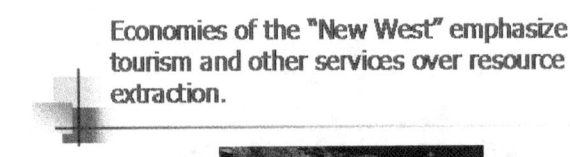

Economies of the "New West" emphasize tourism and other services over resource extraction.

Slide 21

Slide 22

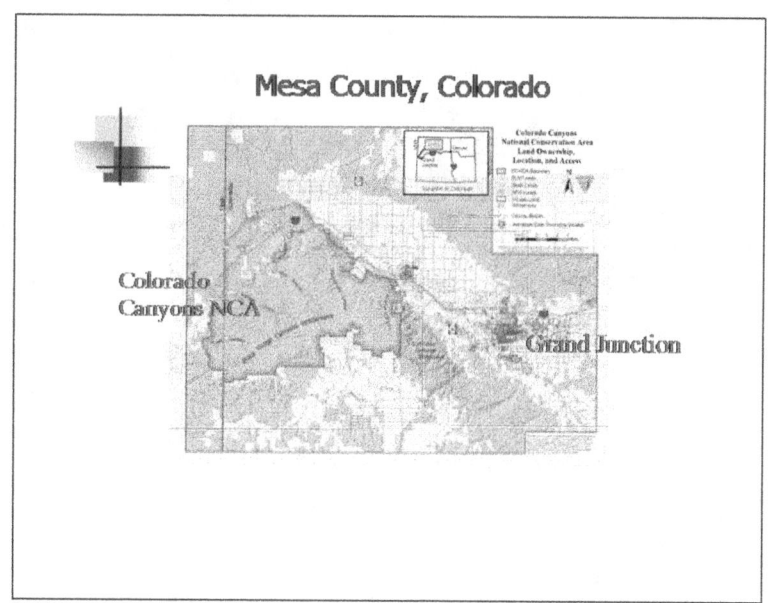

95

Slide 23

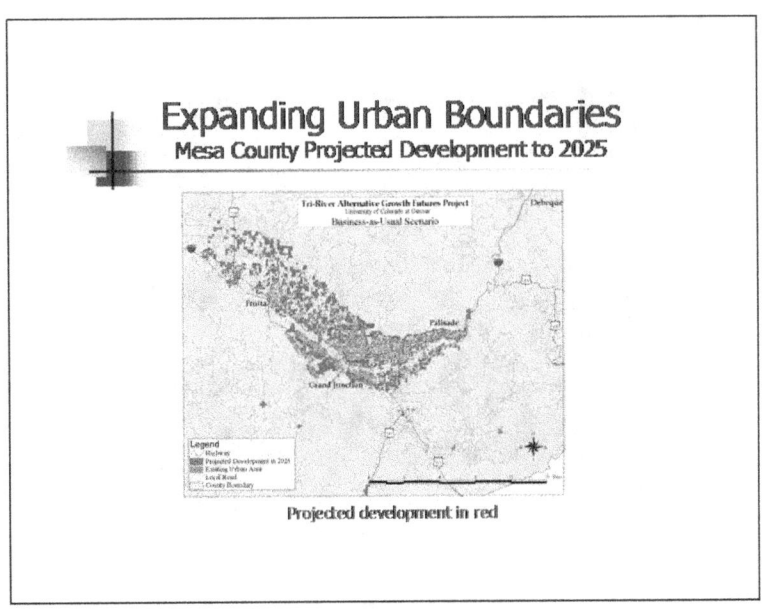

Slide 24

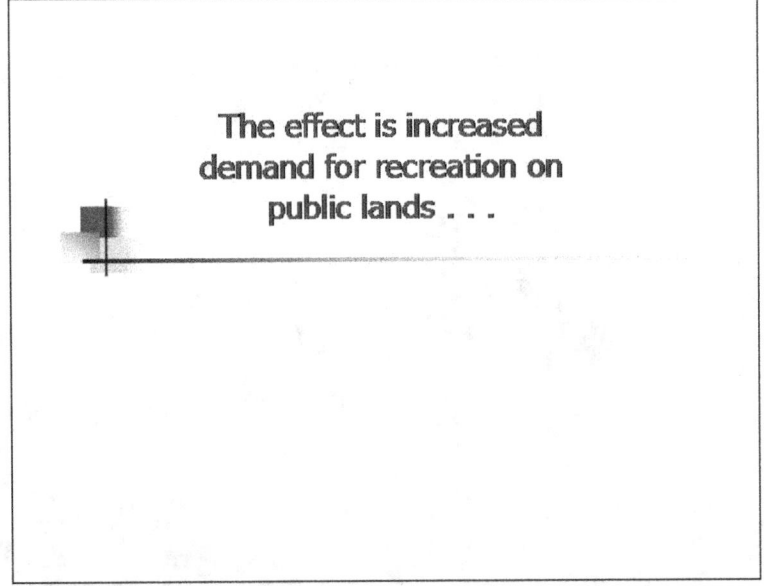

Slide 25

Slide 26

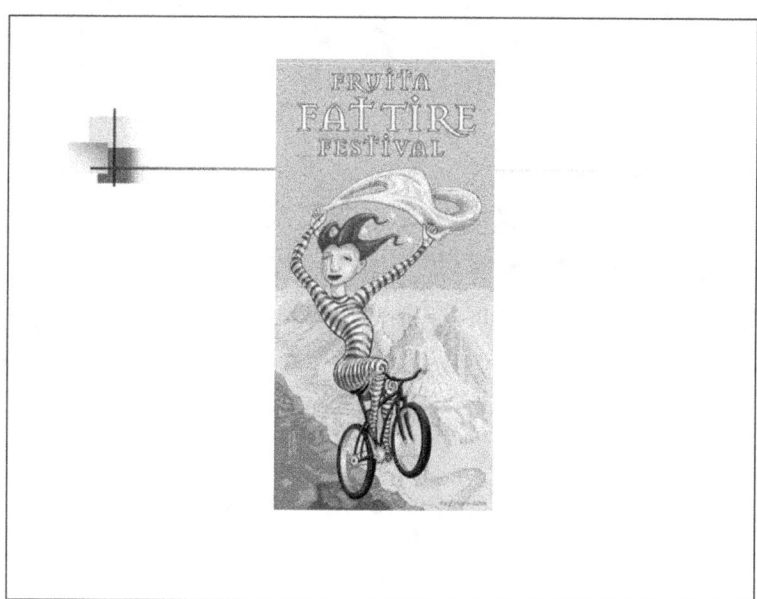

97

Slide 27

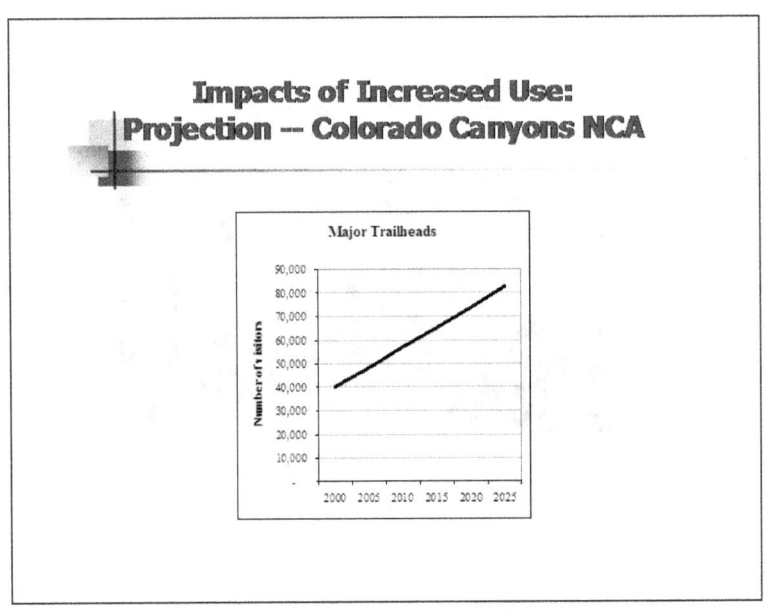

Slide 28

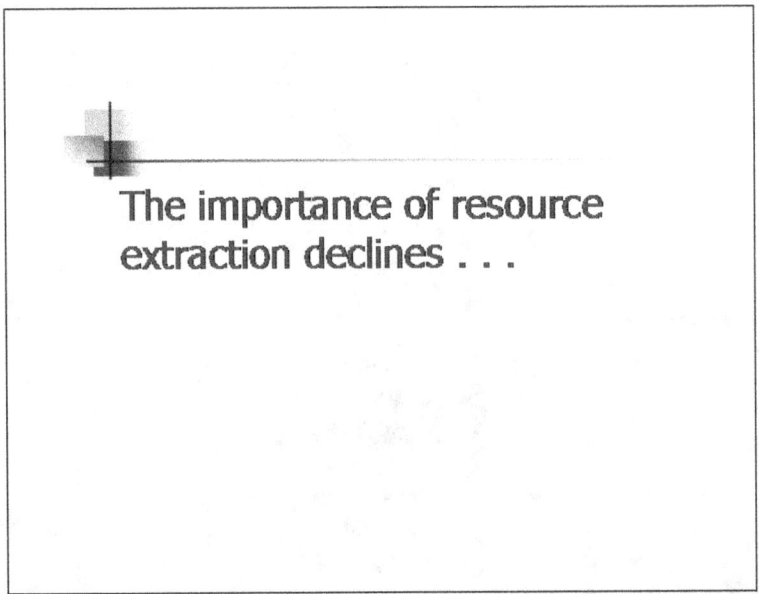

Slide 29

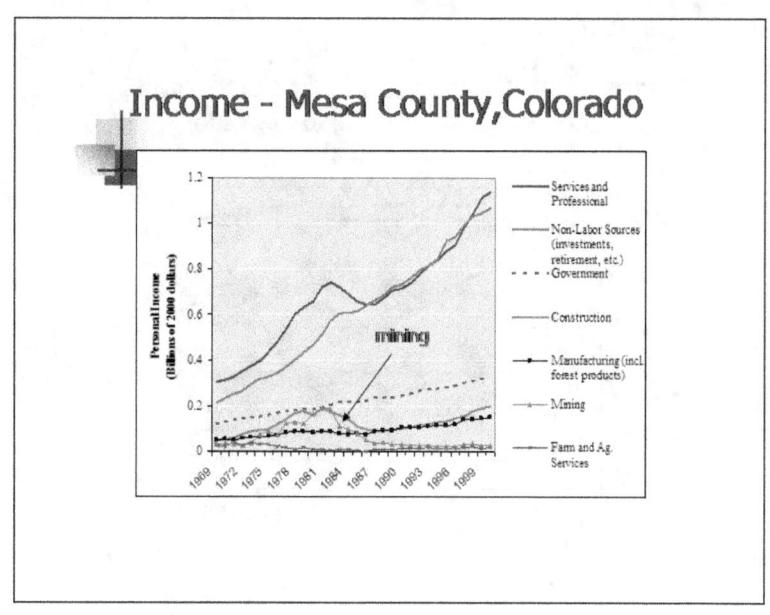

Slide 30

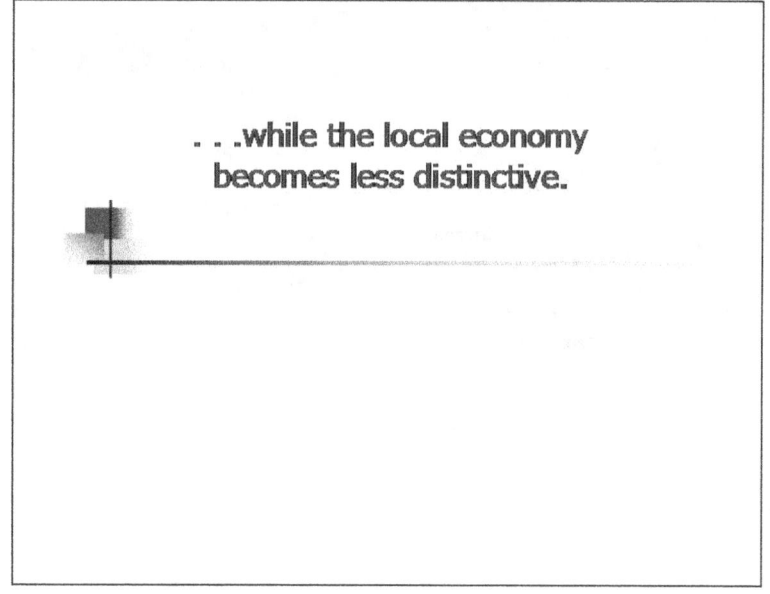

Slide 31

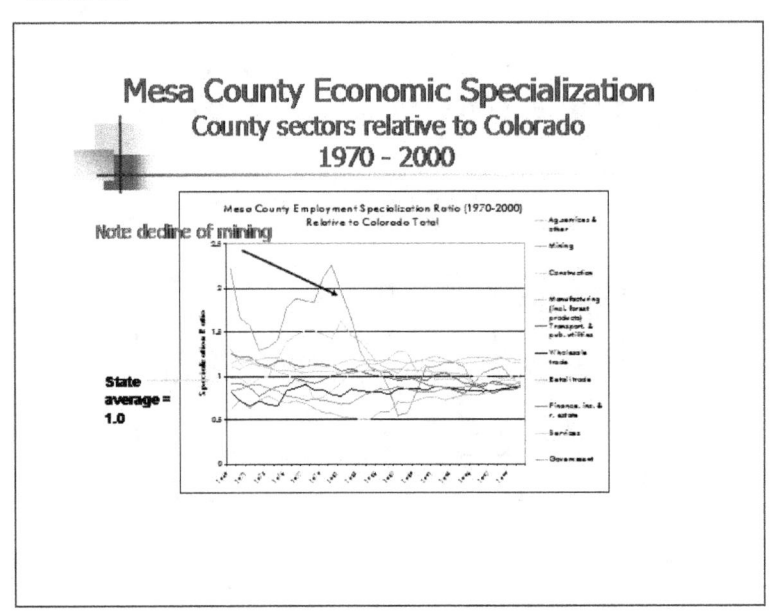

Slide 32

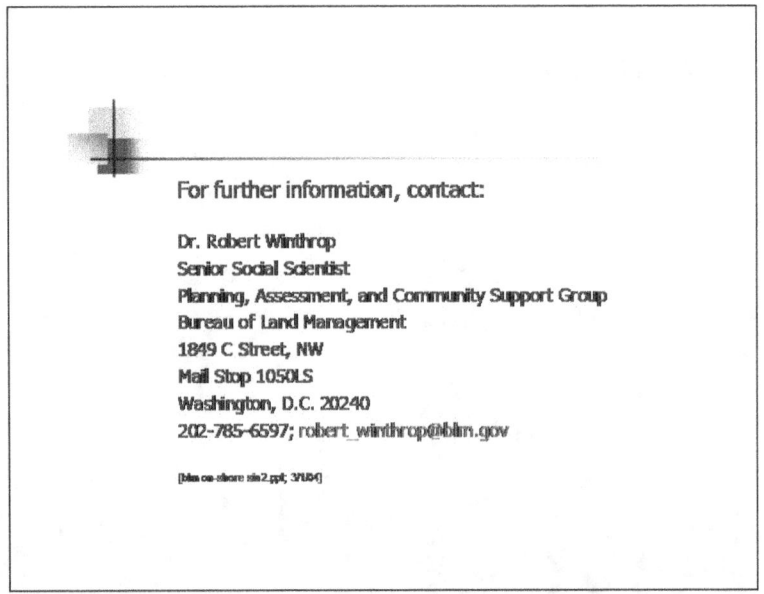

APPENDIX D4

SOCIAL IMPACT ASSESSMENT AND OFFSHORE OIL AND GAS IN THE GULF OF MEXICO

HARRY LUTON AND RODNEY E. CLUCK

Introduction

The National Environmental Policy Act (NEPA) created mechanisms to identify, assess, and mitigate adverse environmental and socioeconomic consequences of government land actions. NEPA-related procedures have evolved into many types of interactions among decision makers and stakeholders, and many forms of information gathering, assessment, and reporting. We address only part of this picture: environmental impact statement (EIS) assessments of socioeconomic effects. Under socioeconomics, EISs evaluate economic, fiscal, demographic, infrastructural, and other social effects such as changes in crime rates and family structure, as well as psychological issues such as the fear of crime or changes in environmental attitudes. We consider only the former types of effects and focus on the logic of NEPA-related assessments rather than on EIS findings. The Gulf of Mexico offshore petroleum industry is our example. Impact assessment of offshore lease sales may entail unique problems, but it also faces the types of general energy-related issues that first motivated NEPA. Thus, we describe here a case in point of how a particular application of NEPA procedures to an ongoing policy in a complex situation raises issues that are common to many current types of social impact assessment.

This paper discusses an "offshore petroleum industry" that is actually composed of many large and small industrial sectors, such as drilling, production, shipbuilding, fabrication, pipelaying, diving and underwater construction, seismic surveying and analysis, trucking, air transport, offshore vessel transport, catering, oil refining, natural gas transport, and petrochemicals. Each of these sectors has its own organization, regulations, dynamics, technological developments, range of labor needs, working conditions, and responses to industry cycles and transformations. As a recent study notes, "Production platforms, once in place, can continue to produce through short-term price fluctuations. Dive companies can find a niche in platform decommissioning, which is sensitive to the age of structures, not the supply and demand for oil and gas. Fabricators, depending on their size and location, may be awarded large and lengthy contracts for deepwater projects, other yards may suffer as smaller development programs are put on hold. Drillers, however, cannot drill unless the owners of leases … initiate exploration and development (E&D) programs" (McGuire and Gardner 2003:219). Current high natural gas prices are good for drillers but devastating to petrochemical businesses that rely on gas for fuel and feedstock. This paper uses such terms as "offshore industry" and "oil industry" to signify this large, diversely sectored, and variably context-responsive set of industries. This paper provides a model for explaining the effects of the offshore petroleum industry on the Gulf of Mexico Region.

101

Background

The Gulf of Mexico offshore petroleum industry is huge. Since the 1950s, more than 5,500 platforms have been installed in the Gulf. Currently, there are approximately 2,400 active platforms on the federal Outer Continental Shelf (OCS), which are the source of 25% of the gas and 30% of the oil produced in the United States. These percentages are expected to rise over the next few years. There are more than 30,000 miles of pipeline on the federal OCS. Approximately 1.6 million miles of seismic lines have been taken on it. There are over 3.5 million passenger trips to and from platforms every year. In 2000, there were 162 shipbuilding and repair facilities in the Gulf and 1,155 registered oil support vessels, 86% of which were based in Louisiana. Also in that year, oil and gas extraction, pipelines, and refining employed over 65,500 people and paid $3.5 billion in wages (Louis Berger Group Inc. 2004). Since 1952, there have been over 90 federal lease sales, and there are currently over 7,500 active leases. The treasury has received approximately $135 billion from bonuses, rents, and royalties since 1982 (see Figure 1).

The Minerals Management Service (MMS) was founded in 1982, created from parts of several governmental bodies overseeing OCS activities. The MMS is charged with leasing offshore petroleum reserves on the federal OCS in an environmentally safe manner. The Environmental

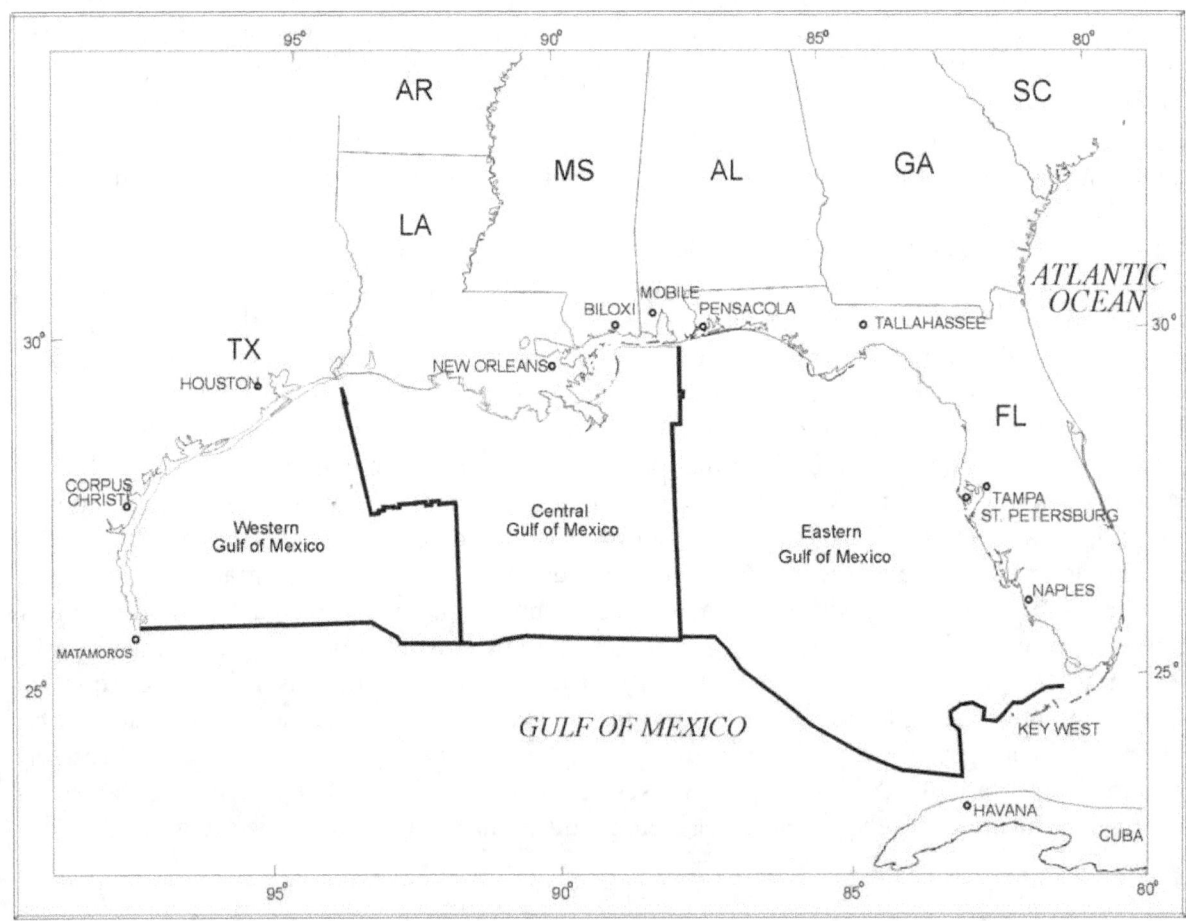

Figure 1: Gulf of Mexico Region.

102

Studies Program (ESP) was originally created under the Bureau of Land Management (BLM) to provide information and analysis in support of agency decision making and environmental assessments as mandated by the Outer Continental Shelf Lands Act. Under MMS, the ESP continues to fund a substantial quantity of research designed to improve the information on which decisions and assessments are made. Since 1982, the agency has invested over $650 million on more than 900 studies of oceanography, ecology, and socioeconomics. However, under the BLM, the ESP conducted little socioeconomic research in the Gulf. The ESP, like NEPA, arose from the ferment of the 1970s energy crisis. The federal response to that crisis included proposals for massive developments of domestic energy sources, which precipitated an effort by federal resource management and science agencies to assess and mitigate the effects of those proposals.

NEPA defines effects as the differences between an area's baseline conditions and the conditions after project initiation. Socioeconomic effects were seen as arising from four sources: (1) national and regional economic inputs; (2) physical and biological impacts of new facility sitings and new sources of pollution, including oil spills; (3) boomtown effects from new labor demand; and (4) public perceptions and fears about these proposals (Pikul and Rabin 1974; NAS 1978). Pointing to the oil industry's long history in the Gulf and its well-developed infrastructure and workforce, the BLM reasoned that regional economic impacts were certain but that any additional social impacts of offshore development would be difficult to identify by means of NEPA baseline-monitoring techniques. As a result, few socioeconomic studies were initiated in the Gulf despite the fact that virtually all U.S. offshore oil development was occurring in that region.

At first, MMS continued the BLM's narrow reading of NEPA and sponsored little socioeconomic research in the Gulf. Then, in 1986, petroleum prices collapsed, sending Louisiana and Texas into recession and convincing many in MMS that the socioeconomic effects of its leasing program were important regardless of their original reading of NEPA. In the 1990s, MMS contracted with the National Research Council (NRC) to examine the adequacy of ESP support for agency decision making and assessments. The NRC was highly critical of the socioeconomic component of the Gulf's ESP, and asserted a strong rationale for conducting research in this area. It noted that the same 100-year history of industry operations in the Gulf that had been used to argue against socioeconomic research also makes the Gulf a ready-made "laboratory" for studying petroleum's social and economic effect (NRC 1992). The NRC reasoned that because the Gulf offshore industry is homegrown, long-lived, and widespread, and includes a complete range of related upstream and downstream activities, most social or economic impacts that the industry can have are likely to be present there. At the same time that the NRC was issuing its report, the MMS was holding its "Gulf of Mexico Socioeconomic Agenda Setting Workshop." The price collapse provided the impetus for the program, the NRC provided its justification, and the 1992 workshop began to define its content (Gramling and Laska 1993; see Luton and Austin 2004 for a discussion of these developments in the ESP).

Since 1992, the socioeconomic component of the Gulf's ESP has grown substantially in size and scope. Over 40 research reports are completed; about 25 projects are ongoing, and more will begin in 2004 and beyond. This paper, however, does not describe the program's growth or content. Instead, it addresses issues that arose as MMS followed the NRC rationale and applied social impact assessment (SIA) logic to the Gulf's offshore oil industry.[1] This paper is divided into four sections. The first discusses what we term the "classic SIA model," the underlying issues, questions, and logic that shape most energy-related socioeconomic assessments, including MMS's. The second uses demography to illustrate differences between the paradigm and the Gulf. What we say builds on other work (e.g., Wilkinson et al. 1982; Gramling and Brabant 1986), and provides a fresh perspective. The third section describes a larger set of problems associated with applying the classic model to the Gulf. The paper concludes with a few comments on where MMS might go from here.

Classic SIA

SIA has evolved in many directions over the years. However, we label as "classic SIA" the model that emerged from a group of impact studies conducted in the 1970s and early 1980s that addressed large, government-regulated projects such as coal-fired generating plants, strip mines, and hydroelectric dams, mostly in rural areas of the western United States (Murdock et al. 1984; for examples of early methodologies, cf.: Wolf 1974; McEvoy 1977; Murdock and Leistritz 1979; Finsterbusch 1980; Finsterbusch and Wolf 1981; Leistritz and Murdock 1981; and Weber and Howell 1982). While this model is often called the boomtown model, we label it classic because it was the first SIA model, the root from which later versions grew, and because it established an underlying logic, set of goals, and list of concerns that still resonate in SIAs that later emerged. Its longevity also makes it classic; for 20 years it has remained a frequently used approach and the predominate model for energy-related projects (e.g., Gilmore and Coddington 1981; Chase and Leistritz 1982; Summers and Selvik 1982; Walsh 1985; Guilford 1989; Ringholz 1989; Vanclay and Bronstein 1995; Luke et al. 2002). As such, it is the approach on which the BLM and MMS began to build their assessments of the social and economic effects of the OCS leasing program (e.g., CEQ 1974; Williams and Horn 1979; Executive Resource Associates Inc. 1984).

As is often noted, this SIA model reflects the conditions from which it emerged, most notably, concerns about boomtowns (Wilkinson et al. 1982; Albrecht 1982). As contemporaries wrote, studies of Gillette and Rock Springs, Wyoming (Kohrs 1974; Gilmore and Duff 1975; Brown 1977; Lovejoy 1977), and, later, of Forsyth and Colstrip, Montana (Blevins et al. 1974; Gold 1974), Craig, Colorado (McKeown and Lantz 1977), Page, Arizona (Ives and Schulze 1976; Little 1977), and Fairbanks, Alaska (Dixon 1978) rejuvenated interest in boomtowns. The basic theme of this literature "is that rapid population growth associated with energy development creates social disruptions, cultural conflict, and pathological behaviors among residents of boomtowns" (Summers and Branch 1982:34–35). These community studies, and others that followed, express a deep concern for a way of life being forever altered—they ask, "Whatever happened to Fairbanks?" They raise issues about local control, about townspeople having little say regarding new forms of industry appearing "on their village green" (Summers and Selvik 1982:vii), about the rural towns becoming dependent on "extra-local decision-making organizations" (Summers and Branch 1982:24), about the idea that mineral exploitation introduces organizational forms that demand changes to the existing social organization, about land-use conflicts and competition for scarce water, about the benefits and uncertainties created for local businesses, about new demands on public infrastructure and services, and about taxes and fiscal policy. They raise issues about large but short-lived labor demand, about the inflow of workers not rooted in the community, and about increasing crime, alcoholism, drug abuse, mental illness, divorce, social isolation, and alienation (Summers and Branch 1982).

Classic Equation: Jobs = People = Effects

While concern about the unique features of communities energized these early impact studies, the recognition of shared characteristics and situations shaped the SIA model. Large, managerially complex, technologically sophisticated, industrial projects designed to exploit a natural resource were constructed near small, rural, isolated, homogeneous, often declining, agriculture-based communities. Project-related decision making was external to the community and foreign to its systems of leadership or governance. Project-related technology, goods and services needs, and labor demands were well beyond the community's capacity to supply. Projects were, essentially, foreign transplants that would exist for a limited time period, divisible into three distinct phases. A short-lived construction phase with high levels of employment and heavy demands on the community and its infrastructure and leadership was followed by a longer operations phase that used fewer, more specialized workers and made lighter, more constant demands. The final phase, decommissioning, brought the project to its inevitable close.

In the classic SIA model, new project labor demand is the primary cause of positive and negative socioeconomic impacts. New jobs increase household income and expenditures, which stimulate local business activity and generate more government revenue through taxes and fees. While higher labor demand always means positive economic effects, the model's strength comes from locating these benefits in their rural context. Because these communities are small and agricultural and the projects are large, industrial, and foreign; because the local labor force lacks the size and skills necessary to fill most of the new jobs; and because these communities are far from any population centers, new jobs will not be filled by commuters already living nearby. New workers must come from elsewhere. This means that inserting large projects into rural communities will cause rapid population growth as outsiders arrive seeking work. Other factors amplify this process. Rural environs compress the new population in space because services like water, electricity, and schools are found only in towns, limiting where in-migrants might live. In addition, the short lifespan of the projects and, especially, of their construction phase compress the new population in time. These communities will experience rapid population growth because new labor is drawn from elsewhere and demand is front-loaded. For the same reasons, these communities will also face rapid population decline. Construction workers who in-migrate will out-migrate when their jobs end, although operations-phase employment may slow this decline. Again, other factors accentuate the ups and downs. Many workers arrive with families, inflating the new population's size and diversity. Since the new households stimulate local business growth, local businesses may also outrun their local labor supply, creating additional in-migration. These jobs, too, will be lost as the project-related population decreases.

An emphasis on links between rapid population change and other effects, such as the fate of local business, is a characteristic of this model that recapitulates its boomtown genesis. While demographic change concentrated in time and space would be an important impact in its own right, in the classic SIA model it is fundamental: it is the first and most germane cause of a wide range of other socioeconomic effects, and it is the point where impact theory, assessment practice, and the original boomtown concerns all meet. As one pioneer practitioner warns, "Determining demographic effects of project development is one of the most important steps in the socioeconomic assessment process because estimating demographic impacts is essential for assessing other population-related effects such as public service demands and fiscal impacts. In fact, too many planners and decisionmakers assume the magnitude of population impacts is synonymous with the magnitude of all impacts" (Leistritz 1992:212).

Socioeconomic Impacts and Their Causes

The classic SIA model addresses several categories of impacts. Demographic effects come first, as products of new labor demand. Economic impacts come second, as products of labor demand and the demographic effects. As already noted, economic impacts amplify the demographic effects (e.g., via secondary or tertiary demand). Infrastructure and public service effects come next. These can include new demands for private and public housing, and for infrastructure and services associated with education, police, fire and emergency services, transportation, water, sewer and sanitation, health and social services, criminal justice, recreation, and libraries. In classic SIA, these effects are due primarily to demographic changes, although some interactions are seen as more complicated. For example, rapid in-migration may create housing booms that increase the tax base along with demands for roads, schools, and police protection. Conversely, the bust brings empty housing, a shrinking tax base, overbuilt schools, and lingering bonded indebtedness.

This housing example raises a fourth category: fiscal impacts. Fiscal impacts are products of project activities, labor demand, and demographic effects. On the positive side, these impacts include increased local revenues (e.g., fees and property and sales taxes). On the negative side, they include increased expenditures to meet new infrastructure and public service demands. Again, this focus on housing, public infrastructure, roads, schools, social services, and public safety parallels concerns endemic to boomtown literature, concerns that also shape classic SIA's basic instrumental goals. Both boomtown literature and the SIA model share the concern that the community and its leadership might be overwhelmed by its swelling population and burdened afterward by an overly optimistic response to it. For this reason, the goal of classic SIA is to produce assessments akin to city planning documents that can be used by affected communities and other relevant jurisdictions (e.g., school districts) to balance responses to the opportunities and difficulties of the economic boom against the realities of the inevitable bust.

The last category generally addressed is social and cultural impacts, which include such topics as the distribution of effects within the community (e.g., who benefits and who is burdened), impacts to specific populations (e.g., effects of inflation on the elderly, the alienation of youth, the isolation of trailer park life), impacts on community cohesion or identification, effects on crime and other dysfunctional behaviors (both actual rates and fear), and effects on environmental attitudes. While some effects in this category may be positive, such as the introduction of new ideas or the increase of what Summers and Branch (1982:39) call "perceived freedom" (e.g., the introduction of alternative pathways to social status), most of them are negative. Social and cultural impacts actually constitute a residual category comprising a variety of topics that share few methodological or subject-area similarities, and that are sometimes considered in an assessment and sometimes not. However, these topics do share one commonality; none fits easily into other classic SIA impact categories because each has a complex and, often, unclear or uncertain relationship to demographic change, a defining characteristic of the other impact categories.[2]

The fact that central boomtown issues, even ones that remain public concerns, are relegated to a residual category illustrates the importance of demographics in defining classic SIA's relevant questions, information, and procedures. However, it also underscores the influence that the model's boomtown origins continue to exert from within. The topic of crime is a case in point. While violent crime is a cause célèbre in boomtown literature, its validity as a contemporary impact has been argued for decades. The debate's very robustness in the face of inconclusive evidence and its marginality to the assessment process may indicate the enduring force of boomtown concerns. More telling, though, is its habitual invocation of Durkheim and anomie, a concept he coined to address social alienation appearing as a rural peasantry was being uprooted into cities that had yet to develop modern structures of social control. In the 100 years since Durkheim's *Suicide*, criminologists have developed alternative explanations of crime rates that are simpler and more directly related to what is known about crime and criminals. Anomie's charm lies not in its simplicity, or elegance, or obvious empirical might but rather in its recourse to an organic community and its breakdown, and to the disorder that must surely follow, and to our gnawing concerns about boomtowns and the worlds we have lost (Summers and Branch 1982; Wilkenson et al. 1982).

Classic SIA as Paradigm

We have labeled the boomtown model as classic SIA because it is the root from which other approaches have sprung. It also fits Thomas Kuhn's famous formulation of a paradigm. Kuhn (1970:10) describes scientific paradigms as models that organize "law, theory, application and instrumentation" into "coherent traditions of scientific research." We identify the classic SIA model as paradigmatic to underscore its importance in shaping theory, application, and technique into just such a coherent tradition. We have already discussed its role in organizing original boomtown concerns, a theory of effects, and the practices and goals of assessment all around the demographic impacts of a project's labor demand. We have also noted that the model defines assessment's salient questions, hence, the salient evidence. The hierarchy of impact categories shaped by their relationships to demography is one example; the issues surrounding criminal behavior are another. Definitions of salience push inquiry in some directions and not others. Studies focus on population-induced demand and, almost in passing, note that petroleum's big infrastructural effect in North Dakota was road wear from truck traffic (Chase and Leistritz 1982). Studies assume a causal nexus between social disruption and a fear of crime most evident in women, while failing to consider a sizable body of literature on the macho, male-centered culture of oil drillers (e.g., Affleck and Eakes 1976; Moen 1986). Finally, we note that, as in

Kuhn's definition of paradigm, this model has shaped SIA's instrumentation. For example, the development of regional input-output (I/O) models is one of the notable achievements under classic SIA (Jones et al. 1988). These I/O models tend to be "static," that is, they assume that relationships among different economic sectors remain constant over time. This assumption fits well with a scenario in which a massive, short-lived project is inserted into a small, rural, agricultural economy. Most goods and services demanded by the project will be imported from elsewhere. However, most contemporary communities are situated in more dynamic economies. Static models are less appropriate when local labor and local enterprises will respond to new project demands.

We also raise the issue of classic SIA as a paradigm to emphasize the large shadow it casts on the entire field of social impact assessment—a wide-ranging but variable influence that is manifested, for example, in the impact categories addressed and their processional order, in an emphasis on demographic effects whatever their magnitude and significance, in the ad hoc and residual character of social and cultural effects even when these are noteworthy public concerns, in the use of fear and anomie and in the general dearth of causal explanations except those rooted in demography or subjectivity, in the focus on communities and local areas and their dynamics almost to the exclusion of larger contexts, and in the focus on projects, the early phases of projects, and construction rather than on the effects of the subject industry (e.g., electric power industry). However, considering this point is not our goal here. Whether one accepts or rejects such family resemblance as evidence of the classic model's wide-ranging influence on the current variants of SIA, it does still seem to dominate the assessment of natural resource extraction and energy projects, and its influence can certainly be found in assessments of the OCS leasing program. We will note several of the many examples of its influence drawn from MMS-sponsored research in the Gulf of Mexico. First, demography tends to be emphasized even when there are virtually no population effects. The Mobile, Alabama, area hosts a large urban and suburban population and complex economy, just the type of context in which many of the OCS program's socioeconomic effects occur. An excellent and detailed study of Alabama's offshore gas industry carefully reports its annual demographic impacts to the tenth of a person, even though these numbers are only artifacts of an economic projection and even though the projected in-migration is inconsequential (Wade et al. 1999). Similarly, a report on the rapid growth of offshore support activity at Port Fourchon was delayed as its author searched in vain for demographic effects that he simply knew must be substantial but, in fact, were hardly there (Hughes et al. 2002). Second, analysis sometimes directly equates demographic impacts with social ones. Not heeding Leistritz's warning, an MMS study of the social costs of the five-year OCS Leasing Program argues that, since the program has no population effects, it has no infrastructural costs (Plater and Wade 2001). These are subtle examples of the influence of the classic model. A spate of research funded by MMS immediately in response to the oil price collapse used the boomtown assumptions explicitly (e.g., Laska 1993; Seydlitz et al. 1993).

Demographics, Offshore Oil, and Classic SIA

The classic approach to SIA grew out of rural Americans' sudden introduction to large, new, energy projects. This approach morphed concerns about small towns and boomtowns into a systematic analysis of socioeconomic impacts. This model may be valid under conditions like those from which it arose. However, like any paradigm, the classic model is a very strong lens that throws the world into a particular focus. Our question is whether this focus is appropriate for viewing the OCS program's social and economic effects in the Gulf of Mexico. We have examined the demographic assumptions that lie at the model's heart. Next, we look at the petroleum industry's demographic effects on Louisiana, particularly from 1960 through the 1980s, to show the marked differences between Gulf realities and the model's paradigmatic assumptions. While the model assumes that affected communities go through similar patterns of in- and out-migration, from early on the demography of south Louisiana petroleum-affected communities exhibited a pattern of stability and geographic differentiation very unlike this scenario. Second, while the model assumes that this pattern will be localized in affected communities, by the end of the 1960s, the Gulf oil industry's demographic consequences were sufficiently dispersed to be systemic; that is, the industry influenced population patterns across the state. Finally, while the model assumes that boomtowns are a consequence of the sudden importation of industrialized processes into nonindustrialized communities, in the late 1970s when southern Louisiana towns boomed, and in the mid-1980s when they busted, these effects occurred in highly industrialized communities with long and strong links to a regional, oil-centered economy (Gramling and Freudenburg 1990a).

Petroleum and Community Effects

The early oil industry certainly contributed to our images of boomtowns. Its Pennsylvania birth gave us the thriving town of Pithole overnight, which now exists only as memory and a roadside marker (Darrah 1972; Gramling 1996). The 1900 discovery at Spindletop and discoveries that followed were marked by a progression of boomtowns—Jennings, Beaumont, Oil City, Vivian, Smackover, and others (Franks and Lambert: 1982). Kilgore, Texas, provoked a classic piece of boomtown literature (Chambers 1933). For the first two decades, discoveries tended to follow the same general pattern. Flush production, the rush to capture oil, and cheek-to-jowl derricks generated booms and busts in nearby towns (Bertrand 1952; Franks and Lambert 1982). However, even during these decades, changes afoot were weakening the forces feeding boomtown growth. Within the industry, technological advances reduced labor demands and lengthened field life while managerial and legal changes tended to stabilize production and the workforce. Outside the industry, ongoing industrialization and specialization, growing regional populations and economies, and improving transportation systems lowered local labor demand, increased locally available labor, and created alternatives for people seeking work while the transformation of U.S. bureaucracies tended to make population movements more manageable (Stein 1964). In oral histories, early industry participants identified Kilgore as possibly the last "real boomtown" (Boatright and Owens 1982), and research on Permian Basin oil booms presents a similar view. It finds that short-lived boomtowns were the exception after World War II. Most places experienced long-term benefits, growing with the boom and then declining, but not to pre-boom levels (Olien and Olien 1982). Postwar America has had its share of booms and busts, but they have been mostly in cities and suburbs, and their causes and consequences are viewed as multidimensional, complex, and, for better or worse, indicative of the country's future rather than as an assault on its past. The 1970s and 1980s phenomena addressed by classic SIA were striking because they moved against the flow of this rapidly urbanizing world, because government energy policies brought a new kind of growth to rural communities that had been in decline for a half century or so.

The petroleum industry that began to develop in Louisiana's coastal wetlands in 1900 and that moved offshore in 1947 was subject to the same trends affecting the rest of the industry. During its formative years and into the 1930s and 1940s, it, too, stimulated the in-migration of "Texicans" and "Americans" and led to boomtown conditions (e.g., Golden Meadow). However, its need to operate in wetlands and over water also made it different. One difference is that oil workers and their families could not live near the fields. Workers had to commute from *terra firma* to their jobs, and the time and costs involved led to a system of concentrated work schedules. Men worked 12-hour shifts for 7, 14, or 21 days straight and then had an equal number of days off. Concentrated work schedules affected the industry's demographic outcomes in two ways. First, they stabilized the residences of the workforce associated with oil exploration and development. While onshore seismic crews, drillers and pipeliners, and their families moved from field to field, in the offshore industry these workers could live in one community and be transported to the various fields. Thus, forces within the industry encouraging a more settled workforce arose earlier and more strongly in south Louisiana than in the industry in general (Austin et al. 2002). Concentrated work schedules also encouraged geographic dispersal of offshore worker residences. Since workers commuted only once every 2, 4, or 6 weeks, they could live relatively far from their point of embarkation (Aratame and Singelmann 2002; Gramling 1980a). This meant that "occupational communities" and occupational segregation found elsewhere in the industry were not as notable in south Louisiana (Affleck and Eakes 1976). Worker households were dispersed within communities, and workers could more easily participate in such "traditional" activities as trapping and fishing (Gramling 1983). This pattern of dispersed worker settlement patterns associated with the maintenance of more rural lifestyles and long commutes to work to nonurban job sites reflects a general southern pattern of industrialization that differed from the earlier northern norm because it occurred in the age of electric power and as road systems began to improve. In the south, industrialization was not synonymous with urbanization. Factories were built outside of cities, and workers settled in more rural areas such as "ribbon" developments along connecting roads. By the 1940s, this pattern was "particularly pronounced" in basic industries in south Louisiana, an area where the petroleum and petrochemical industries and rural communities had already had almost a half-century of interactions (Heberle 1948:34).

Operations in wetlands and over water also required a much larger, more complex, and industrialized support structure than was needed by the rest of the industry to build, operate, and maintain the platforms, exploratory rigs, seismic vessels, pipelaying barges, and various supply and support boats required. It also made many more demands for goods and services generally. Thus, on one hand, compared to its onshore relative, the offshore industry involves many more sectors of the economy and requires a much larger and more diverse labor force. On the other, the offshore industry served to stabilize the residences of this worker population. In classic SIA, the construction workers move from job to job, thus creating the boom phase of projects. Offshore oil development requires much more front-loaded labor than do onshore projects. However, while this labor force may be analogous to construction workers in classic SIA, fabrication yards, shipyards, and ports are geographically fixed, and their workforce lives nearby, weakening any tendencies to create boomtowns, at least as envisioned by classic SIA. This analysis also highlights a second aspect of the infrastructure needed to work offshore that, in the long run, proved to be even more significant in shaping the industry's demographic effects—its enormous size and complexity.

To summarize, even early on, oil industry operations in wetlands and over water led to a larger, more diverse, and more settled workforce in southern Louisiana than was typical for the industry at large. Also, early on, these differences led to community-level demographic consequences that differed markedly from the classic SIA boomtown scenario. The 1900 discovery at Jennings, on the heels of Spindletop, and the discoveries that followed at Vinton, Walsh, Iowa, Hackberry, and other communities were marked by a rush to production and rapidly increasing and decreasing populations. The Jennings salt dome, however, has now been producing for over 100 years. It exemplifies the changes in technology and strategy that have led to more stability within the industry in general (see Forbes 1946). More relevant here are the changes specifically related to operating in wetlands and over water. Already by the 1920s, several larger southern Louisiana towns such as Morgan City, Lafayette, and Lake Charles had become oil supply centers, providing labor, services, and fabricated equipment to oil fields over a wide territory, and some companies also located their main administrative centers near the action. As this diversification occurred, their population growth became more robust and showed a diminished sensitivity to drilling activities in nearby fields. This was a process of industrial aggregation (Heberle 1948). These communities were strategically located to serve the oil industry because of their railroad connections and their access to water transport, and because they were already serving as centers for trade and manufacturing. Also, by the 1920s, several smaller but strategically located southern Louisiana communities such as Golden Meadow were already serving as bedroom communities for rig workers commuting to platforms located in the area's uninhabitable wetlands (Bertrand 1952). Even before classic SIA scenario had been written, the dynamics of the southern Louisiana oil industry were undermining its validity.

Petroleum and Systemic Effects

In 1947, drilling moved from near shore into the open ocean, and as the offshore industry grew, so too did its needs for onshore support. Within two decades, it had transformed itself into the economic motor behind Louisiana's expanding economy (Scott 1978), so much so that, even after its downturn in the 1980s, about 3,800 contractors and vendors in 47 parishes were still providing about $4 billion in goods and services to OCS operations (Applied Technology Research Corp. 1994).

Local specialization and its demographic effects were already evident in the 1920s (Bertrand 1952). By the 1960s, the size and spread of the industry and its enormous demands for goods and services affected economic opportunities and demographic outcomes statewide. These effects appeared against the backdrop of a declining, traditional, agricultural-based, Southern economy. Except for three decades, Louisiana experienced net out-migration from the Civil War until the 1970s. This means, more people left the state than moved to it except when the recessions and

111

depressions of the 1890s, 1900s, and 1930s limited economic opportunities elsewhere. This relocation from agrarian rural areas to more industrialized urban centers disproportionately affected Louisiana's blacks, who have shown net out-migration for every post-Civil War decade except the 1870s (Maruggi and Wartenberg 1996). These differences indicate differences in opportunities. During the decade of the 1960s, as southern Louisiana's oil industry began to grow, the state's pattern of out-migration began to change. There was still a steady net out-migration from rural areas and blacks continued to leave at higher rates than they entered, but the net migration rate for whites became slightly positive. Moreover, the growing presence of the oil industry and its role in making the state more economically attractive to white immigrants are evident in the geographic distribution of the shifts. In the 1960s, 13 of Louisiana's 64 parishes experienced net in-migration and all but two of these were in the urbanized, industrialized, and oil-influenced southern half of the state. One exception was due to a military buildup at Fort Polk; the other was a northern urban center (Christou 1972; Maruggi and Wartenberg 1996). In Louisiana's southern half, parishes experienced net in-migration due to a white flight to suburbia and to real economic growth in the New Orleans and Baton Rouge metropolitan areas and the industrial strip stretching between the two. This real economic growth was driven in large measure by growth of the oil industry and associated refining and petrochemical industries. Also telegraphing the industry's emerging regional demographic effects was the growth of the coastal parishes of St. Mary and Lafourche, which were heavily involved with supporting its offshore activities (Burford and Murzyn 1972). Thus, by the end of the 1960s, the oil industry's demographic consequences had ceased to be the kind of localized or community-centered phenomena addressed by classic SIA. Its imprint was found on the growth of cities and their suburbs, on the growth of industrialized regions, and on the general pattern of growth and decline throughout the state of Louisiana.

While the petroleum industry is known for its volatility, the 1960s began a decades-long period of a generally upward movement as the country's demand for oil and gas grew. This upward movement accelerated in the 1970s, particularly after the Arab oil embargo and U.S. policy responses to the crisis, and continued to accelerate into the 1980s (Baxter 1993; Wallace et al. 2003). During the 1970s, offshore production outpaced that of onshore (Lindstedt et al. 1991), and this powerful economic force and the myriad opportunities it was generating were strongly evidenced in the patterns of demographic change throughout the state (Scott 1981). The 1970s was the first decade since the great European immigration of the 1870s in which Louisiana experienced substantial net in-migration; 32 parishes—half of all parishes—showed net in-migration, and even black out-migration slowed. This growth occurred against the backdrop of a still-declining rural Southern economy. While all areas of the state performed well, continued weakness in the northeast and along the Mississippi "indicate that agriculture and forestry based economies in the 1970's fared worse than the petroleum and manufacturing based areas" (Maruggi and Wartenberg 1996:39, 41). Also, certain national demographic trends were more in evidence than in previous decades. For example, as elsewhere in the country, suburbanization was a dominant factor in 1970s population growth. However, net in-migration was clearly related to the oil boom and the accelerated job growth that began in the early 1970s, and that accelerated after 1974 with the Arab oil embargo and the federal policy responses to it (Maruggi and Saussy 1985). The shift in Louisiana's patterns of demographic growth during this era was a product of forces affecting economics at many levels. First, this shift still should be seen against the backdrop of a declining, traditional, agriculture-based economy for it was due to increased in-migration rather than to decreased out-migration. Also, this shift should be seen in the light of the national economy, for it was pushed along by considerably slower employment growth elsewhere (Maruggi and Wartenberg 1996). Finally, it should be seen in the light of wider national and international conditions that helped deepen the country's recession and drive up the price of oil to levels that were to prove untenable (Baxter 1993).

Net in-migration continued to accelerate into the early 1980s until oil prices began to fall and, then, to drop dramatically. By 1986, the oil industry had gone bust. Again, the industry's demographic effects were felt throughout the state. In the 1980s, only four parishes experienced net in-migration, and these resulted from the continued trend toward suburbanization, not from economic growth. St. Tammany had the only significant amount of net in-migration, and that was due to an equally significant out-migration from New Orleans. In the 1980s, 60 parishes and all metropolitan areas experienced net out-migration. As in the preceding decade, the demographic consequences of the industry were statewide and should be viewed against the backdrop of other regional and national trends. Again, the cause of net out-migration was not as much from an increase of people leaving Louisiana so much as from a virtual cessation of people moving into it. Again, the rate of in-migration was strongly influenced by the national economy. "The Louisiana economy again moved in the opposite direction of the national economy in the 1980s. Oil prices plummeted in 1981, triggering the oil-bust economic recession that lasted through most of the decade. In contrast, the United States enjoyed solid economic growth in nearly every year of the 1980s. Thus, the loss of the high-paying oil patch jobs that attracted workers to Louisiana in the 1970s resulted in an unprecedented out-migration of 411,099 persons in the 1980s" (Maruggi and Wartenberg 1996:11).

The industry's internal dynamics (e.g., technology intensiveness), coupled with its unique mix of economic and geopolitical issues (e.g., elasticity of demand, OPEC influence on supply), make it more volatile than many, and one might assess how its unpredictability impacts states, communities, and individuals. Clearly, the ramping up of offshore activities in the 1960s and 1970s, and their spike and dive in the 1970s and 1980s, had marked demographic outcomes that overlaid and intertwined with the other demographic trends that defined these decades—the decline of rural southern agriculture and suburbanization, for example. Just as clearly, the classic SIA model sheds little light on these salient industry effects. The model simply cannot illuminate because of its narrow geographic and temporal focus; because of its emphasis on the construction phase of projects to the neglect of other activities, agents, and processes; and because of its attempts to explain the effects of the offshore petroleum industry without recourse to an understanding of that industry's organization, scale, geography, and rhythms.

In the heady days of the late 1970s and early 1980s, times were flush, and the national news was quick to identify south Louisiana boomtowns with their flocks of jobseekers living in labor camps and cars. Then, during the traumatic mid-1980s oil crash, the same media reported on people fleeing, businesses failing, and houses being repossessed (e.g., Trillin 1979; Schweid 1989).[3] Classic SIA arose from the empirical grounding of case studies of project-induced boomtowns. Below, we examine Louisiana's reported oil boomtowns, focusing on Morgan City as the most notorious example, to show that these events differ sharply from the SIA model and are best explained as local manifestations of larger, industry-wide changes.

Louisiana Boomtowns circa 1980

The offshore industry is large, complex, and unevenly distributed among communities. As we have seen, this differentiation was evident by the 1920s. Oil-involved towns were becoming supply bases, equipment fabricators, bedroom communities, refining centers, banking and service centers, or combinations of these. By the 1960s, the oil industry had also grown sufficiently large to have demographic consequences across the state. These two trends are obviously related. As the oil industry grew, so too did its role in the state and regional economies. Also, as its economic role grew, so too did the importance of its uneven distribution in shaping its local-level social and economic effects. This issue of distribution involves both quantity and mix. With regard to quantity, when offshore exploration increases, all involved communities should experience increases in local business activities, but communities with many oil-related businesses should benefit more than those with few. The local mix of oil-involved businesses is important in shaping benefits because the offshore industry is composed of many different types of businesses, each of which responds differently to any given change. Therefore, if offshore exploration and development increases significantly, communities that host a mix that is heavily involved in drilling or new platform fabrication should benefit more than those with a mix primarily related to production or refining. The opposite is also true. Should offshore exploration decline, communities most involved with exploration and development should suffer most.

This interplay between offshore exploration and development and the local industry mix highlights the differences between the role of labor demand and demographic change in classic SIA and their actual relationship in the Gulf. In both, early project phases generate high levels of blue-collar labor demand, perhaps more so in the Gulf since labor-intensive activities associated with drilling and fabrication form large sectors of offshore industry employment and are in highest demand during exploration and development. Blue-collar labor is generally more sensitive to industrial fluctuations than is white-collar. Thus, in classic SIA, tenure is short for construction workers and long for production workers. Similarly for the offshore industry, its large blue-collar workforce is concentrated in such activities as fabrication and drilling, all of which are in high demand during exploration and development but not during production. Exploration- and development-related jobs are particularly sensitive to offshore ups and downs because, compared to production, the costs of these activities are higher and more discretionary, and their returns less immediate and predictable. In classic SIA, the workforce is external to the impacted community, and project phases are discontinuous; hence, these phases are expressed as the comings and goings of workers. In the Gulf, discontinuous phases also employ an enormous blue-collar workforce. However, unlike that in classic SIA, this workforce is already resident in the affected communities. Moreover, while offshore projects are discontinuous, they do not produce discontinuities in local labor demand onshore for two basic reasons. First, oil-involved communities serve multiple projects in various phases in multiple fields of varying maturity. Thus, communities experience the labor demands of various projects and project phases as a

114

blend. Second, the separation of labor into phases is anything but clear-cut. Fabrication and drilling are strongly associated with exploration and development, but fabricated ships support production platforms and need refitting, exploration and maintenance drilling occurs on production platforms, and most workers are welders, crane operators, crewmen, or are in other non-phase-specific jobs. Southern Louisiana has a higher percentage of jobs in mining, transportation, and fabrication than the U.S. average (Manuel 1980; 1997), but the workers look like workers anywhere; most live at home, commute to work, and work onshore. For communities, their residents, and the local mix of businesses, the demands from exploration and development are experienced as part of the general demands of the offshore industry for goods and services and not as identifiable phases. Under most conditions, changes in the level of these activities are experienced as normal ups and downs in the local demand for goods and services. As in many U.S. industries, in good times, jobs segue into the next; in bad times, workers may be laid off.

Morgan City, in St. Mary Parish, like other southern Louisiana boomtowns of the 1980s, is an example of the interplay between local industry mix and offshore exploration and development under extreme conditions. Morgan City, particularly, was a lightning rod for these dynamics because of its concentration of the most blue-collar, labor-intensive, and activity-sensitive businesses in the industry (Gramling and Freudenburg 1990b; Manual 1985). This evolution began as a lumber town with transportation advantages. Morgan City emerged in the 1850s as a residential and commercial area because of its railroad connections to New Orleans and steamer service to Galveston (Broussard 1977). Its first boom came when the northern investor Charles Morgan built a port and dredged a channel to the ocean to avoid the unionized facilities in New Orleans. By the early 1900s, it was already an important inland port hosting commercial fishing and shipbuilding industries (Baughman 1968). This pattern of aggregation typifies industrial growth in rural Louisiana; new industries were attracted to established ones because of shared transportation, infrastructure, and labor needs (Heberle 1948). In Morgan City, oil was a part of this aggregation. The town caught oil fever in 1901 even before a well was produced in the state (Broussard 1977), but it became an operations center for the coastal wetlands because it was a port and transportation hub with a shipbuilding industry, a blue-collar labor force, and access to the oil fields (Gramling 1984a; Davis 1990). Later events reinforced these advantages. The 1905 development of the Intracoastal Waterway extended the port's inland reach (Franz and Durio 1977). The 1935 discovery of the Jeanerette field put Morgan City in the center of the state's top petroleum-producing parish. A 1930s offshore shrimping boom also added to the port's transportation and shipbuilding capacities. World War II brought more shipbuilding and another boom, further solidifying the town's identity as an oil and fabrication center. Then, in 1947, Kerr-McGee completed a well on nearby Ship Shoal, often considered the first successful offshore well. At that moment, Morgan City already was a supply base and fabrication center and was poised to meet the emerging demands of the offshore industry and to grow as that industry grew.

Morgan City's 150-year life as a port and industrial center includes several periods of rapid growth and three with reported housing shortages which all occurred in a wider social context that made the problem look similar to ones in other places. In the early 1900s, its bustling port attracted a "floating population" of Norwegians, Swedes, Poles, Greeks, Italians, Portuguese, and

Filipinos (Baughman 1968). A wave of European immigration in the 1900s put similar but larger foreign populations in such southern port cities as New Orleans and Galveston (Maruggi and Wartenberg 1996; Donato 2004). During WWII, its shipbuilding industry attracted depression-displaced people seeking war-related marine construction work (Broussard 1977). The war began a demographic revolution as sharecroppers fled rural depression for the rapidly industrializing cities (Daniel 1990). Morgan City's 1980s boomtown experience follows a similar path. As noted, the city was positioned to benefit from the development of the offshore industry. Increased job opportunities in the area "brought a rapidly growing population and increased permanent settlement particularly during the 1950s and 1960s. Entire communities such as Bayou Vista sprang into being" (Gramling 1984b:134). By the mid-1970s, Morgan City lay at the center of an enormous collection of industry-related enterprises. Traveling east from the town was "a seemingly never-ending galaxy of industries, including the sprawling McDermott and Avondale Shipyards, which employed literally thousands of men and women from throughout St. Mary, and even as far as St. Martin Parish." For miles there was "an immense diversity of industries: shipyards, oil field helicopter operations, supply houses, and a plethora of other businesses all catering to the needs of the gargantuan oil companies" (Broussard 1977). Thus, Morgan City grew rapidly throughout the 1950s, 1960s, and 1970s as part of a much larger offshore-oriented industrial expansion. However, despite its concentration of development-stimulated businesses, at no time during this economic boom did it, or similar communities, become a boomtown. Gramling (1983) examines this period and finds that Morgan City did not because the local evolution of the offshore industry and the pace of population growth gave the community time to adjust (see also Gramling and Brabant 1986). While he found that offshore growth kept Morgan City's property values and rents relatively high from the 1940s into the 1980s, any housing shortage in the immediate environs of Morgan City was due to land availability problems because of the surrounding wetlands and competition from agriculture, and it was mitigated by longer commutes to work. He also argues that, during the 1960s and 1970s, the industry's high wages allowed south Louisianans to achieve a middle-class lifestyle (Gramling 1980b, 1984a; Manuel 1997).

Despite three decades of sometimes rapid growth in the industry, the surrounding parishes, and the town, Morgan City achieved its supposed boom- and bust-town status only within the first few years of the 1980s, and this status again reflected current events elsewhere. The "boom" came after the Arab embargo, the superheated oil industry, a credit crunch from the savings-and-loan collapse, and a deep recession devastating much of the rest of the country's industries, and after newspaper hype about plentiful jobs and easy wealth in the oil patch sent recession-displaced workers from across the country flocking southward, sometimes with their families and sometimes alone and with only what their cars could carry. Many of the earlier arrivals did find jobs, but insufficient work for all placed great burdens on shelters, soup kitchens, and local helping institutions in general (Brabant 1993). But even these problems may have been somewhat overstated, described as they were in the immediate aftermath of the oil-price crash and in the light of the boomtown model. For example, the presence of camps for offshore oil-related support workers was used as evidence of a boomtown housing shortage in the 1980s (Brabant 1993). However, while a housing shortage may have existed and the labor camps no doubt grew larger as the industry boomed, these camps exist in good times and bad for they serve day-labor recruited for the lowest paid types of oil-related work. They are analogous to migrant

116

labor housing, not indicators of local housing availability (Higgins 1999).[4] Then, world oil prices began to fall precipitously and, for a time, exploration and development virtually ceased. The oil crash devastated the economies of Oklahoma, Texas, and Louisiana, but communities like Morgan City were most hurt. By the mid-1980s, businesses were closing, workers were losing jobs or taking pay cuts, and people were leaving, mostly newer arrivals but also locally-born residents. What once had been viewed as a vehicle carrying blue-collar workers into the middle class was now a community's "overadaption" to an inherently cyclic extractive economy (Gramling and Freudenburg 1990b).

Morgan City is no more a classic boomtown than is Flint, Michigan, which suffered through plant closures when the regionally dominant automobile industry reorganized in the face of Japanese competition. The 1980s oil-price crash came at the end of a decades-long expansion of a massive industry that extended from Texas to Alabama, after OPEC actions had heated that expansion to a boiling point, and after a growing recession elsewhere in the country set droves of laid-off workers south to find work. These events were not the result of the completion of a project or a group of projects, or of happenings in Morgan City, but rather occurred because the town lay at the heart of a regionally dominant industry suffering a regionwide economic depression that rolled through Louisiana, Texas, and Oklahoma as oil prices collapsed and exploration almost stopped. Actually, the 1980s oil price bust inverts the causal relationships postulated by the classic SIA. Effects occurred because the industry labor demand was long term and widespread and was not compressed in time and was not within only one or a few communities. Out-migration was not the cause of this downturn; rather, it occurred as oil's downturn brought down other sectors of the economy, and it was countercyclic, stimulated by economic expansion elsewhere in the country. Similarly, social services were overloaded because of a shrinking state tax base, not because of local demand. Causes were manifestations of larger-scale processes and of exceptional combinations of oil- and non-oil-related factors, such as an unusual concentration of fabrication jobs, an offshore industry overheated by an OPEC embargo, an economic recession elsewhere sending workers south to seek jobs, and a city growth limited by wetlands.

Demography Modeled; the Gulf Summarized

The demographic consequences of Louisiana's petroleum industry contradict the core assumptions of the classic SIA model—that new project labor demand causes demographic change that causes other project-related socioeconomic effects. The 100 years of experience with oil operations in coastal wetlands and the 50 years of experience with offshore operations mean that the ongoing development and operation of existing offshore projects, the initiation of new ones, and a labor force poised to meet the demands of both are part of Gulf baseline conditions. The industry does not appear in communities as something new with discrete phases but rather as a continuation of business, and social and economic effects relate to changes in the magnitude and mix of this commercial continuity.

Morgan City during the crash is an extreme example of an extreme event since its concentration of exploration- and development-oriented industries was unusual even for the oil patch. During the bust, nearby Abbeville performed better despite its similar dependence on offshore work, due in part to a greater proportion of this work being in the less-volatile production side of the industry. Similarly, New Iberia's better performance and quicker recovery were helped by its more white-collar mix of oil-related industries and by its role as a bedroom community for Lafayette (Tolbert 1995; Tolbert and Beggs 2004). While the bust staggered all Louisiana, significant differences in community-level effects were generated by differences in the local configurations of oil-related businesses—that is, by each community's oil industry mix.

The crash was an historic upheaval, but the relationship it exposes between the industry, its local industrial mix, and its social and economic effects holds true generally. Simply stated, change in the offshore oil industry affects each of its sectors differently. A change in the offshore industry is expressed in each involved community through changes in its oil-involved businesses. Since each community hosts a different mix of oil-related businesses, each is differently affected by a change. One caveat must be made—the local conditions that amplify and mitigate effects also differ among communities. As a contemporary example of this process, the ongoing restructuring of the oil industry is concentrating white-collar jobs in Houston. This advantages Houston and disadvantages New Orleans regardless of the ups and downs in overall white-collar employment levels, and its demographic impacts are more concentrated in the New Orleans middle-class suburb of Mandeville than in blue-collar Morgan City, regardless of the latter's greater dependence on the offshore industry.

To conclude, as the industry grew into a driver of the state's economy, the local specialization already apparent by the 1920s became a template by which the larger industry wrote its effects on involved communities. In principle, any change to the industry should have locally differing effects. Oil-involved communities may be tied to the industry's fate, but each is tied to it in a very specific way. Thus, while the classic SIA model begins with a community and a new project that will generate a certain magnitude and pattern of demographic effects, Gulf realities begin outside of the community, and by linking community-level outcomes to larger industry changes, these realities seriously complicate the assessment of demographic effects. Below, we compare and contrast classic SIA and the Gulf to illustrate these complications.

Figure 2a is a generic representation of classic SIA and a Gulf model. In it, a change at A generates effects at B, C, and D, and a larger change at A means larger effects at D. The hexagram A is the initiating cause. In both models, the B-C interface is where economic inputs transform into socioeconomic effects. In this schematic, D is the "community in general." It is represented by a trapezoid to emphasize that its existing conditions shape the effects of A. Both models assume that these conditions at D are products of past and current social and economic trends, and both call these conditions the "baseline," here represented by the circle "base." Both models view baseline conditions as exogenous variables that may entail opportunity costs.

Figures 2b and 2c represent classic SIA and the Gulf, respectively. In Figure 2b, a new project (A) creates new labor demand (B), which generates demographic effects (C), which produce other socioeconomic effects within a community (D). The initiating cause (A) triggering this chain of effects is a NEPA-related event such as a new generating plant. The B-C economic-to-social interface translates labor demand (B) into population change (C). Relevance for baseline

118

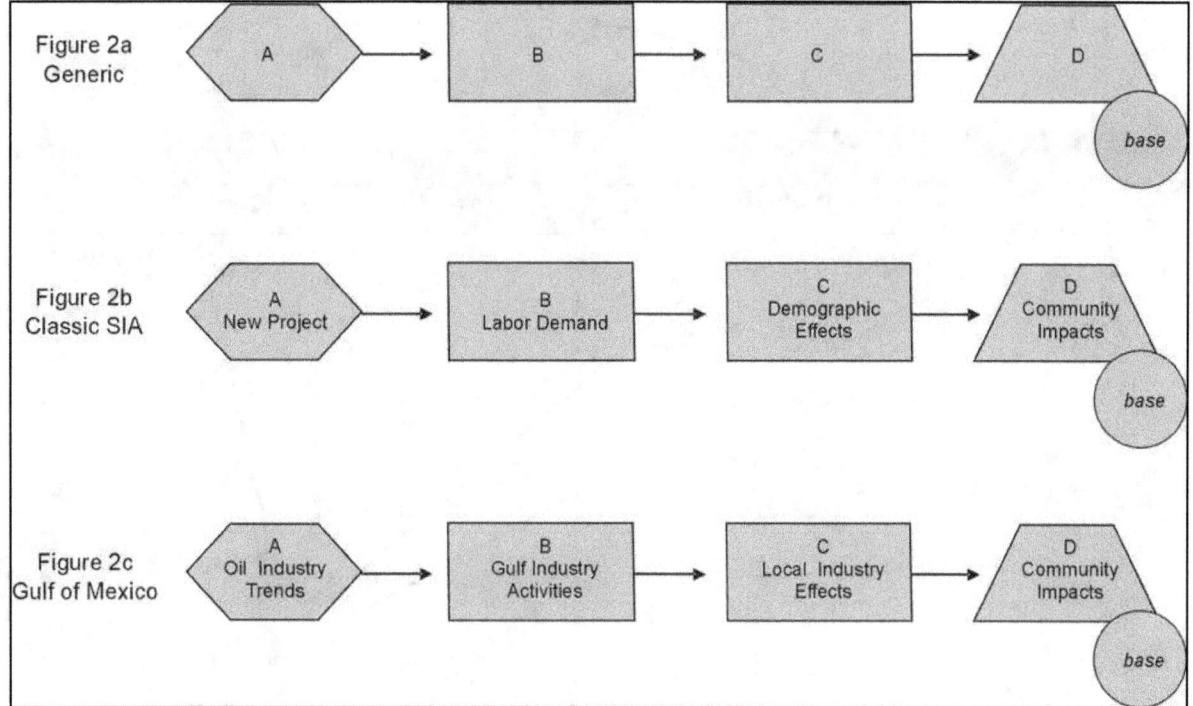

Figure 2: General model.

conditions at D is determined by the content of this interface. For example, in classic SIA, a large labor force at D is relevant because it limits the effect of A by reducing in-migration at C.

In Figure 2c, oil industry trends and events (A) affect Gulf oil activities (B), which impact the local oil industry mix (C), which leads to other socioeconomic effects within a community (D). In this model, the initiating cause (A) may or may not relate to NEPA-triggering events. Thus, compared to Figure 2b, the initiating cause in the Gulf is less easily specified, understood, or linked to government actions. In Figure 2b, demography is not a determining part of the causal chain. Instead, these effects are included as one of many possible community-level outcomes at D. In the Gulf, the B-C interface translates changes in industry organization, strategies, technologies, and demand (B) into changes in local business practices (C). Again, the B-C interface is difficult to specify, a condition mirrored in the wide range of relevant baseline conditions. For example, in some Gulf communities, the sugar industry buffers petroleum's fluctuations. There, an assessment of the future of federal sugar supports is relevant (Tobin 2001; Wallace et al. 2001).

Figure 3 further develops the Gulf model and the issues of scale, timeframe, and causal complexity that render classic SIA inappropriate. While classic SIA is designed for a project in a

community and resides entirely within that community, Gulf effects originate outside the community and cannot be explained by community-level events. In the Gulf, petroleum industry trends and events (A) are geographically unspecified; impacts on the Gulf's petroleum industry (B) can originate anywhere. In Figure 3, B lies within the line labeled "region," which includes

119

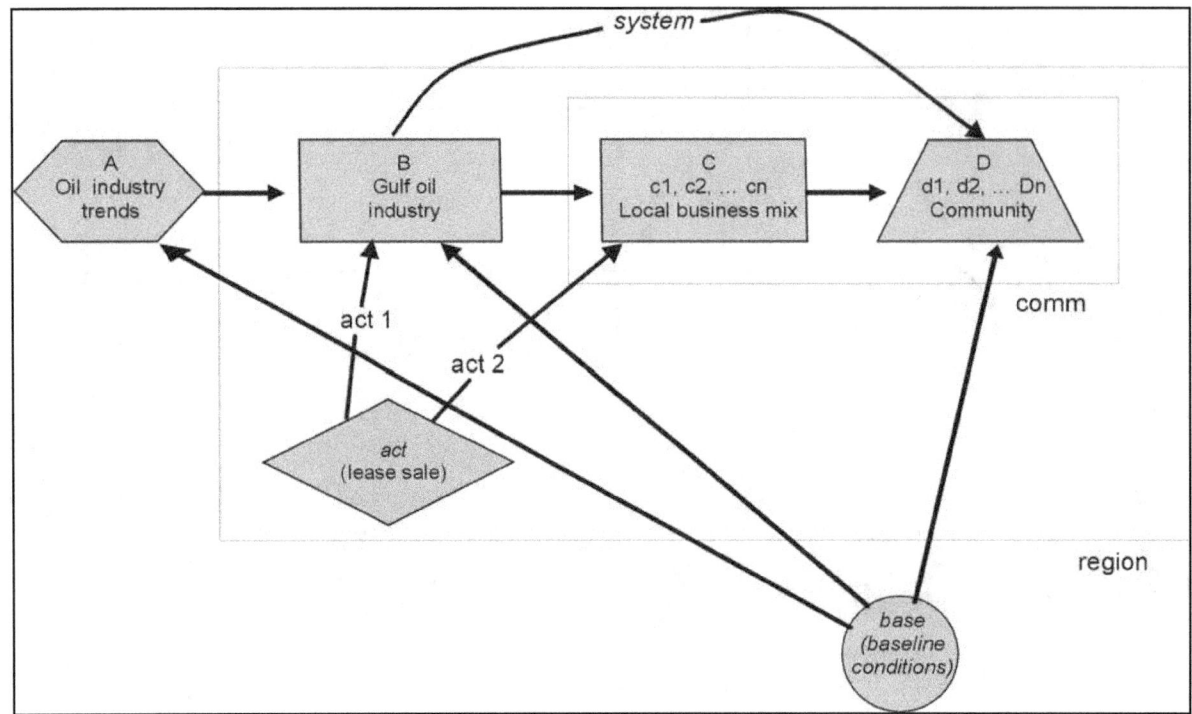

Figure 3: Gulf model.

parts or all of Texas, Louisiana, Mississippi, Alabama, and Florida. In this model, only C and D reside in the community. In Figure 3, the line "comm" designates one or several jurisdictions, towns, or parishes.

The Gulf's extensive geography makes problems for data acquisition and analysis but more intractable ones regarding causation. In classic SIA, specifying relationships within its causal structure is straightforward. Labor demand of planned project A minus available labor in community D equals unmet labor demand at B. Unmet labor demand times average new laborer household size equals in-migration at C. In the Gulf, a change in OPEC production at A will affect industry behavior at B and C, which can have demographic effects at D, but identifying these casual links and their values is neither straightforward nor certain. Difficulties multiply for C and D. The five-state Gulf Region includes hundreds of counties, parishes, communities, and relevant jurisdictions (e.g., school districts). In Figure 3, multiple industry mixes are designated as c1, c2, … cn and multiple communities as d1, d2, … dn. Even if the effect of A on B can be estimated, the problems of distributing this value among the numerous industry mixes and of estimating its effects on the associated communities remain.[5]

Timeframes raise other difficulties. Classic SIA assesses projects and effects with relatively short and predictable life spans; indeed, a key characteristic of demographic effects is temporal compression. In classic SIA, only community trends and conditions (D) have long life spans, and these complications are part of its baseline. In the Gulf, timeframes are generally long with indeterminate beginnings and endings. Ramifications of events at A like the oil embargo or the emergence of deepwater exploration reverberate far into an unknown future, as do outcomes at B

120

like the growth of Port Fourchon (Hughes et al. 2002). The lease sales MMS assesses are serial events in an areawide leasing policy that helped shape Louisiana's offshore oil industry over decades (Priest 2003). Long timeframes mean that industry events and trends, the Gulf oil industry, and the community all have relevant baselines, as represented in Figure 3 by the circle "base" with causal arrows pointing to A, B, and D. (Current conditions at C are products of the effects of B as mitigated by D.) These baselines add difficulties to specifying model components, assessing relationships among them, and to distinguishing petroleum's effects from other historical trends as rural decline or suburbanization.

This issue recalls another. In the Gulf, the NEPA-triggering action being assessed plays only a minor role in the industry trends and events that generate offshore oil's onshore effects. Figure 3 represents the NEPA-triggering action (e.g., a lease sale) with the parallelogram labeled "act" which lies within the region but to the side of the main A-D causal chain. A lease sale might amplify or mitigate trends at B and C, including effects of the ongoing OCS program, but a sale is not their primary cause. In Figure 3, arrow act 1 designates region-level lease sale effects. The MMS projects a lease sale's future employment effects based on oil prices and past industry behavior. Arrow act 2 designates local-level lease sale effects. Project sizes and locations become known as companies submit development plans. These are not estimated at the time of the lease sale.

Figure 3 highlights another difficulty with causation. In both models, effects move through a causal chain from A to D. However, in Louisiana, the oil industry looms sufficiently large in the state's economic, tax, and fiscal systems that a change in its fortunes has an effect on all its communities that is independent of, and in addition to, the more direct effects from local industry involvement (Brabant 1994; Scott 1978; 1981). Figure 3 represents this situation with a causal arrow labeled "system" pointing from B to D. System effects also vary geographically because of local conditions and tax and fiscal policies.

A New Approach to Understanding Socioeconomic Impacts in the Gulf

Early BLM planners were not far off when, based on their narrow reading of NEPA, they eschewed consideration of most socioeconomic effects in the Gulf. They were wedded to a model that focused on community-level effects of a government action and that assumed demographic impacts substantially determined other socioeconomic consequences (e.g., CEQ 1974; Pikul and Rabin 1974). Examples of such effects resulting from a specific OCS lease sale are rare indeed. The NRC call to use the Gulf as a "natural laboratory" for the study of the industry's socioeconomic effects was intended to broaden this approach, not to challenge the SIA paradigm. Nevertheless, to use this natural laboratory, the agency had to adopt a more eclectic approach to petroleum's effects, and one that recognized oil's effects are interwoven with other local, regional, and national trends in a "dynamic baseline" (Smith 2000). One unforeseen consequence of this change was that MMS began to confront, head on, previously latent challenges to the application of classic SIA analysis to Gulf lease sales. Five of these challenges are listed below.

121

(1) <u>Challenge of Defining the Affected Area</u>: The first task in socioeconomic assessment is to define the affected area. This raises the question of scale. The area must be large enough to include the significant impacts yet small enough not to dilute them. The MMS assesses the OCS program's socioeconomic effects on the Gulf Region, which includes Texas, Louisiana, Mississippi, Alabama, and Florida. This encompasses all relevant effects but dilutes many. It also raises the problem of variation within the study area. The OCS effects are shaped by each state's fiscal and tax policies, the distribution of other industries, and the industry's own purchasing and hiring patterns (Plater et al. 2000; Luke et al. 2002; Hughes et al. 2002; Dismukes et al. 2003). The Coastal Zone within these states is a smaller affected area, but it, too, includes some effects while diluting others, and its 56 counties and parishes also include the extremes of social, economic, cultural, and institutional variation. The Coastal Zone is comprised of literally hundreds of cities, towns, school districts, port authorities, levee boards, special tax regions, and other tax jurisdictions that separate and join the parishes and counties in different ways. Social and infrastructural effects are often defined by specific local conditions—the unused capacity of a certain school district, the growing demands on a particular water system, or the condition of a specific road connecting a port and highway (Keithley 2001). Defining the boundaries of states and coastal zone counties and communities may be simple, but identifying and describing the salient variation within such wide-ranging "affected areas" is not.

(2) <u>Challenge of the Baseline</u>: Under NEPA, the difference between an area with and without the proposed action is the proposal's effects. The area sans proposal is the "baseline." However, since the industry has operated in the Gulf for decades, there is no "unaffected environment," hence no baseline as originally envisioned by NEPA. While this has led some assessments to conclude that the program has no socioeconomic effects to be addressed, it has led others to ascribe all problems faced by oil-involved Gulf communities to the industry. This tendency is evident in much of the research MMS funded in its immediate response to the 1980s oil price bust, leading one frustrated oil executive to observe that even if southern Louisiana had never had oil, it would not, today, be an untouched Acadiana of happy fisher folk and trappers (Porter 1992; Seydlitz and Laska 1994). The task of separating the effects of oil from other regional influences and from larger national and worldwide trends is neither easy nor certain. Past effects of oil and gas development on communities, families, and individuals are bound up in other "baseline" trends.[6] Many social forces impinge on communities, families, and individuals such as mass communication, changes in education, and increasing community heterogeneity. Even in oil-involved areas, the industry is just one of many causes of most effects (Wallace et al. 2001). Identifying oil's share of an impact is made more difficult because most of its impacts are not unique to that industry (Shrimpton and Storey 2001).

(3) <u>Challenge of Identifying the Offshore Oil Industry</u>: SIA addresses the effects of an offshore petroleum industry which itself lacks clear boundaries. This industry is actually composed of many types of enterprises involved in the processes of finding, extracting, refining, and bringing petroleum-based products to market. Even the basic activity of drilling a well is usually undertaken by a number of firms and individuals interacting through contracts and subcontracts and supported by a myriad of other firms and individuals involved in such activities as legal or

122

insurance work, trucking materials, and providing food. Even in onshore areas where the oil industry is relatively small and where many of its sectors are not present, as in North Dakota, the numbers of enterprises required and the variability in their sizes, organization, and interactions make projecting the effects of onshore oil development extremely difficult (Chase and Leistritz 1982). In the Gulf, the challenge is immeasurably greater because of the size and complexity of this industry, because the full spectrum of enterprises involved in finding, extracting, processing, storing, and bringing petroleum-based products to market is present (Louis Berger Group Inc. 2004), because the support and transportation requirements of offshore operations add substantially to the complexities and variabilities of the oil industry (Manuel 1983; Gramling and Brabant 1984), and because of its uneven distribution across the Gulf. Each type of industry has its own structure, dynamics, technologies, infrastructural requirements, labor organization and demands, and links to communities and the economy, and for each industry, these attributes are changing over time (Wallace et al. 2003). The fabrication, pipe-laying, drilling, diving, trucking, and supply boat industries all face different demands from the industry move to deep water, and they also face different demands and opportunities onshore (Austin and McGuire 2000; Wallace et al. 2001). Finally, the situation is complicated because only portions of many of these industries are oil-related. For example, questions about banking services, trucking, or port activities become particularly difficult since their relationships to the petroleum industry vary from place to place in ways that noticeably affect local socioeconomic outcomes (Tobin 2001; Tolbert 1995; Tolbert and Beggs 2004).

(4) Challenge of Addressing Local Effects: The MMS assesses socioeconomic effects for lease sales. However, its EISs do not address the act of leasing but, rather, the effects of future industry actions that may result from sales. The EISs assume that all industry activity that occurs on a lease is caused by the lease sale, a useful assumption, but not literally true since leases only create opportunities for actions while the actions are undertaken based on economic and business considerations that change over time. Thus, to analyze a sale's potential effects, MMS must develop a scenario that estimates the activities that will occur on leased blocks that is much removed from any actual project plan. Using the scenario, and based on past industry behavior, MMS then projects economic and employment impacts for large, multi-county "subareas." Sale-level scenarios are necessarily general and lack the geographic specificity that would support projecting sale effects at county or community levels. However, many types of social and infrastructural effects only occur in the context of local conditions—the unused capacity of a certain school district, the growing demands on a particular water system, or the condition of a specific road connecting a port and highway. Thus, MMS faces the question of how to relate multi-county projections necessitated by sale-level assessments to an assessment of local-level effects. Just as the baseline challenge implies that onshore effects cannot be linked to specific sales, this challenge questions the possibility of linking effects of a sale to specific onshore locations.

(5) Challenge of Addressing Cumulative Effects: The challenge of cumulative effects relates to the baseline challenge. Since the industry is already in place, a lease sale's primary socioeconomic effect is to continue the status quo by maintaining the arena in which the industry operates. The state of Louisiana has repeatedly complained that sale-level assessments do not adequately address the real effects of the OCS program because its real effects are cumulative.

While the baseline challenge addressed the problem of separating oil's effects from other regional influences and from broader national and worldwide trends, the state of Louisiana raises the issue of whether sale-level effects can be meaningfully separated from ongoing effects and assessed independently when past industry outcomes are actually conditioning the expression of the new ones. For example, one reason boomtowns are unlikely in the Gulf is that the local labor force has already been strongly shaped by the industry. This does not mean that all effects lie in the past but, rather, that the industry's current fluctuations are expressed in such terms as job insecurity rather than by migration (Austin et al. 2002; Donato 2004).

Classic Model Revisited

This list developed piecemeal during attempts to apply classic SIA, yet it raises problems with the entire classic approach, from defining the study area to considering cumulative effects. Gulf problems with the paradigm's demographic heart are part of a wider and more widely relevant disconnect. Below we contrast SIA and the Gulf Region (see Table 1), but the "Gulf Region" could be any agency that must assess social impacts of complex causes in complex situations.

Classic SIA assesses the effects of a project while the Gulf Region assesses the effects of a program. Under classic SIA, projects are relatively simple, and key variables (e.g., type, location, size, and labor demand) are derived from project plans; thus, they are action-related and localized. In the Gulf Region, the petroleum industry is the affecting agent. It is extremely complex, and key variables must be estimated using scenarios based on resource estimates and past industry behavior; thus, key variables are hypothetical and general. Classic SIA addresses projects in one or several communities while the Gulf Region addresses a program in five states. Classic SIA addresses project effects in small, rural, isolated places while the Gulf Region addresses program effects that occur primarily in urban settings or in long-industrialized rural ones. These differences mean that classic SIA confronts processes akin to industrialization or modernization while the Gulf Region considers ones akin to regional growth and decline.

Contrasts in study areas are mirrored in equally striking ones concerning effects. In classic SIA, projects are new and foreign to the impact area while, in the Gulf Region, the oil industry is decades old and familiar. In classic SIA, projects have discontinuous life spans with planned beginnings and foreseeable ends. In the Gulf, even though oil is a nonrenewable resource, industry effects are ongoing and will continue for decades. In classic SIA, the limited lifespan of a project heightens effects by compressing them in time. In the Gulf, activities rise and fall, but industry effects are not compressed in time. In classic SIA, project life spans are experienced as phases (e.g., construction, operations, and decommissioning), each with its own spectrum of effects. While the Gulf Region uses phases to project the industries' economic effects, all onshore activities are ongoing and overlapping and phases are indistinguishable. In classic SIA, community experiences vary by phase, while in the Gulf, they vary by industry mix and activity levels. While classic SIA addresses large-scale energy projects, the compression and segmentation of time lead to a focus on the construction phase and construction trades. In the Gulf, fabrication is one piece of an entire industry that is the focus.

124

Table 1: Differences between the SIA paradigm and Gulf realities.

SIA	GOMR
Assesses a project	Assesses a program
Project simple and key variables specific to plan and geography	Industry complex and key variables hypothetical and general
Assessment area = the community	Assessment area = five states
Community small, rural & isolated	Affected areas include urban and/or industrialized
Processes related to industrialization	Processes related to regional development
Project new to area	Program (and industry) developed in area
Project timeframe discontinuous	Program timeframe ongoing
Time compression highlights effects	Timelines not compressed
Timeframe segmented	Timeframe segments all ongoing & overlapping
Effects vary by project phase	Effects vary by industry mix and activity level
Effects concentrated in construction	Effects of segments indistinct
Project imposed from without	Program (industry) evolved in area
Project organization unfamiliar	Industry tied to local entrepreneurship
Project technology unfamiliar	Project technology locally developed
Project scale massive & unfamiliar	Project scale typical and familiar
Labor demand greater than supply	Local labor supply matched to industry
Labor demand compressed in time	Labor demand continuous
"Boom and bust" concerns	Market fluctuation concerns
Cumulative effects = other projects	Cumulative effects = ongoing program
Effects decision driven	Effects economically driven
Assessment rationalistic	Assessment probabilistic
Outcomes more "predictable"	Outcomes less "predictable"
EIS stresses planning	EIS stresses documentation

In classic SIA, a project is imposed from without, and project organization and technology are unfamiliar, while, in the Gulf Region, the offshore industry developed over decades, and its organizations and technologies are familiar. In classic SIA, this imposition raises issues about the adequacy of local responses and, particularly, about the adequacy of the labor force. In the Gulf Region, communities and their labor forces have adapted to the industry. In Classic SIA, new labor demand compressed in time raises the specter of booms and busts. In the Gulf, new projects do not normally lead to unusual labor demands; instead, fluctuating activity levels affect communities. Reflecting its concern with the new and imposed, classic SIA defines cumulative effects in terms of other foreseeable future projects. In the Gulf, sales are serial events, and cumulative effects are defined in terms of the ongoing leasing program.

In classic SIA, effects flow from the decision to allow a project to proceed, while, in the Gulf, lease sale decisions play a minor role in generating industry effects. This means that the assessment process for classic SIA can be rationalistic and deductive and based on project variables, while, in the Gulf, assessment builds on scenarios and complex industry interactions and must be more inductive and probabilistic. This also means that outcomes of the NEPA-related analysis are more predictable under classic SIA than they are for the Gulf Region. Finally, reflecting differences in familiarity, the meaning of cumulative effects, and predictability, classic SIA stresses the usefulness of assessments as planning documents while Gulf Region stakeholders see their use as official documentation of the OCS program's effects.

Conclusions

The SIA methods discussed here were formulated to measure impacts from single, often one-dimensional causes (e.g., a generating plant), of relatively short duration (e.g., several years), in small and easily definable areas (e.g., communities), where the impacting agent is externally imposed and where this agent overwhelms the community's institutional structures, infrastructural capacities, and labor force. Oil development in the Gulf lies at the opposite end of the continuum for each of these qualities. The industry is immensely varied and complex; it has evolved over generations and in concert with other social, political, economic, and technological changes that often dwarf, mask, inspire, and mitigate the more conspicuously oil-related ones; and it has grown enormously. Its influence has reached across the seas, its end is not in sight, and no final tally of impacts is possible.

Our overview of the industry's demographic effects shows traditional NEPA-style analysis to be woefully inadequate to the task of addressing their scale, complexity, and duration and that the classic SIA paradigm, because of its community- and project-level focus, cannot provide the needed conceptual framework or methodological tools. Moreover, the problems identified in the Gulf are characteristic of many assessment situations faced today. Most impacts are not rural; most impacted communities are not physically or culturally isolated; most impacted economies are not agricultural; and most impacts are not caused by demographic change. Agencies assess ongoing policies and programs, multiple and difficult-to-isolate causes, and varied and complex impact areas. Indeed, as with the thousands-of-year-long timeframes of nuclear waste disposal sites, some projects may fit the classic paradigm but raise issues that do not. The question is, "What is to be done?"

We suggest a strategy for answering this question rather than a new conceptual model. NEPA and classic SIA emerged together, but one does not require the other. Classic SIA's advantage is a coherent analytic framework, built around a NEPA event, that logically relates assessments of one type of effect to the others. Our strategy sacrifices this coherence for a more empirical approach. No single explanatory model should be expected to best suit the varied assessment situations agencies face. In complex situations like the Gulf, a model for assessing state-level impacts would not necessarily be useful for local-level ones. Models, like any form of explanation, should reflect available data and scientific practice.

The same may be said about the topics addressed. Lists of topics are essentially ad hoc, drawn from NEPA scoping, other agency information-gathering efforts, and existing SIA literature.

126

Classic SIA assumes, with little evidence, that assessed topics form an interacting system of effects and causes. We suggest that a strategy begin by addressing each of topic separately, while accepting that the mechanisms by which it is affected, the degree to which it is affected, how these effects relate to others, and whether they merit inclusion in the final assessment are all empirical questions. For each topic, analytic coherence will come from the logic and findings of relevant academic fields—from criminology when looking at crime, for example—and not from the topic's role in an *a priori* model. The classic model defined some relationships as important and others as not. An empirical approach to impact topics may provide a more useful foundation on which to build future monitoring and mitigation efforts.

Topic selection has always recognized differences in assessment situations. Analytical strategies should also reflect such differences. For example, in the Gulf the complexities of the petroleum industry necessitate the agency's very substantial effort to understand it, and its long-lived operations magnify the importance of cumulative effects for the assessment process. Similarly, the need to assess sale-level effects turns the problem of linking regional- and local-level effects into a strategic one.

While this strategy will not draw the picture of tightly linked effects provided by classic SIA, it will produce one that is more complete and is coherent enough. Consistency will come from the need to assess each topic in terms of its role in overall effects and from addressing the peculiarities of the assessment situation—the need to link regional-level effects to local-level ones, for example. While the EIS is the legal vehicle for reporting findings, supplementary reporting methods should also be explored. While classic SIA presents a simple explanatory scheme that can be encompassed between the covers of an EIS, our strategy may produce material inappropriately massive or technical for that setting.

We have described the reasons MMS will continue to pursue a more realistic and robust approach to evaluating social and economic changes associated with the Gulf Region's OCS leasing program. We have also begun to outline a strategy this pursuit might take. To the extent this strategy reflects a shift away from the traditional SIA paradigm, we believe it is well founded and long overdue.

References

Affleck, M. and R. Eakes. 1976. Oklahoma Roughnecks: The Case for an Occupational Community. Free Inquiry 4(2):114–151.

Albrecht, S.L. 1982. Empirical Evidence for Community Disruptions. Pacific Sociological Review 25:297–306.

Applied Technology Research Corporation. 1994. Louisiana, Gulf of Mexico Outer Continental Shelf offshore Oil and Gas Activity: Impacts. Baton Rouge, LA: Mid-Continent Oil and Gas Association.

Aratame, N. and J. Singelmann. 2002. Effect of the Oil and Gas Industry on Commuting and Migration Patterns in Louisiana: 1960 to 1990. New Orleans, LA: U.S. Department of the Interior, Minerals Management Service, Gulf of Mexico OCS Region. OCS Study MMS 2002-072.

Austin, D., K. Coelho, A. Gardner, R. Higgins, T. McGuire, J. Schrag-James, S. Sparks, and L. Stauber. 2002. Social and Economic Impacts of Outer Continental Shelf Activity on Individuals and Families, Volume 1: Final Report; Volume 2: Case Studies of Morgan City and New Iberia, Louisiana. Prepared by the University of Arizona, Bureau of Applied Research in Anthropology. New Orleans, LA: U.S. Department of the Interior, Minerals Management Service, Gulf of Mexico OCS Region. OCS Study MMS 2002-22 and 2002-023.

Austin, D.E. and T.R. McGuire, eds. 2000. Social and Economic Impacts of OCS Activities on Individuals and Families: A Report for the Participating Communities. Tucson, AZ: University of Arizona, Bureau of Applied Research in Anthropology.

Baughman, J. 1968. Charles Morgan and the Development of Southern Transportation. Nashville, TN: Vanderbilt University Press.

Baxter, V.K. 1993. Political Economy of Oil and Exploitation of Offshore Oil. *In* Impact of Offshore Petroleum and Production on the Social Institutions of Coastal Louisiana. Laska, S.B., K.B. Vern, R. Seydlitz, R.E. Thayer, S. Barbant, and D.J. Forsyth, eds. New Orleans, LA: U.S. Department of the Interior, Minerals Management Service, Gulf of Mexico OCS Region. OCS Study MMS 93-0007. Pp. 15–42.

Bertrand, J.A. 1952. Oil and Population in Southern Louisiana: 1901–1935. Master's thesis. Baton Rouge, LA: Department of Geography, Louisiana State University.

Blevins, A.L., J.G. Thompson, and C. Ellis. 1974. Social Impact Analysis of Campbell County, Wyoming. Laramie, WY: Wyoming Environmental Institute.

Boatright, M.C. and W.A. Owens. 1982 [1970]. Tales from the Derrick Floor: A People's History of the Oil Industry. Lincoln: NE: University of Nebraska Press.

Brabant, S. 1993. From Boom to Bust: Community Response to Basic Human Need. *In* Impact of Offshore Petroleum and Production on the Social Institutions of Coastal Louisiana. Laska, S.B., K.B. Vern, R. Seydlitz, R.E. Thayer, S. Barbant, and D.J. Forsyth, eds. New Orleans, LA: U.S. Department of the Interior, Minerals Management Service, Gulf of Mexico OCS Region. OCS Study MMS 93-0007. Pp. 195–208.

Brabant, S. 1994. The Ripple Effect of Resource Development: Ouachita Parish as Case Study. Impact Assessment 12:59–74.

Broussard, B. 1977. A History of St. Mary Parish. Franklin, LA: Self-published.

Brown, B.S. 1977. The Impact of the New Boomtowns: The Lessons of Gillette and the Powder River Basin. Washington, DC: U.S. Government Printing Office.

Burford, R.L. and S.G. Murzyn. 1972. Net Migration for Louisiana and its Parishes, 1960–1970. Division of Research Occasional Paper No. 8. Baton Rouge, LA: Louisiana State University, Division of Research College of Business Administration.

Chambers, W.T. 1933. Kilgore, Texas: An Oil Boom Town. Economic Geography 9:72–84.

Chase, R.A. and F.L. Leistritz. 1982. Socioeconomic Impact Assessment of Onshore Petroleum Activity. Paper presented at Second International Conference on Oil and the Environment, Halifax, Nova Scotia.

Christou, G.C. 1972. Migration in Louisiana, 1960–1970. Research Study No 17. New Orleans, LA: Louisiana State University, College of Business Administration, Division of Business and Economic Research.

Council on Environmental Quality (CEQ). 1974. OCS Oil and Gas: An Environmental Assessment. A Report to the President by the Council on Environmental Quality. 5 vols. Washington, DC: U.S. Government Printing Office.

Daniel, P. 1990. Going Among Strangers: Southern Reactions to World War II. Journal of American History 77 (December):886–911.

Darrah, W.C. 1972. Pithole, the Vanished City. Washington, DC: Library of Congress.

Davis, D.W. 1990. Living on the Edge: Louisiana's Marsh, Estuary and Barrier Island Population. Louisiana Geological Survey 40:147–160.

Dismukes, D.E., W.O. Olatubi, D.V. Mesyanzhinov, and A.G. Pulsipher. 2003. Modeling the Economic Impacts of Offshore Oil and Gas Activities in the Gulf of Mexico: Methods and Applications. New Orleans, LA: U.S. Department of the Interior, Minerals Management Service, Gulf of Mexico OCS Region. OCS Study MMS 2003-018.

Dixon, M. 1978. What Happened to Fairbanks? The Effects of the Trans-Alaska Oil Pipeline in the community of Fairbanks, Alaska. Boulder, CO: Westview Press.

Donato, K.M. 2004. Labor Migration and the Deepwater Industry. New Orleans, LA: U.S. Department of the Interior, Minerals Management Service, Gulf of Mexico OCS Region. OCS Study MMS 2004-057.

Executive Resource Associates, Inc. 1984. Federal Outer Continental Shelf Oil and Gas Activities: A Socioeconomic Review. Detailed data tables, 12 vols. Washington, DC: U.S. Department of the Interior, Minerals Management Service.

Finsterbusch, K. 1980. Understanding Social Impacts: Assessing the Effects of Public Projects. Beverly Hills, CA: Sage.

Finsterbusch, K. and C.P. Wolf, eds. 1981. Methodology of Social Impact Assessment. 2nd ed. Strousburg, PA: Hutchinson Press.

Forbes, G. 1946. Jennings, First Louisiana Salt Dome Pool. Louisiana Historical Quarterly 29(2):496–509.

Franks, K.A. and P.F. Lambert. 1982. Early Louisiana and Arkansas Oil: A Photographic History, 1901–1946. College Station, TX: Texas A&M University Press.

Franz, R. and C. Durio. 1977. Transportation. *In:* E.F. Stallings and T.F. Reilly, eds. Outer Continental Shelf Impacts: Morgan City, Louisiana. Louisiana State Planning Commission, pp. 209–220.

Gilmore, J.S. and D.C. Coddington. 1981. Socioeconomic Impacts of Power Plants. Research Project 1226-4. Denver, CO: University of Denver, Electric Power Research Institute, Industrial Economics Division.

Gilmore, J.S. and M.K. Duff. 1975. Boom Town Growth Management: A Case Study of Rock Springs-Green River, Wyoming. Boulder, CO: Westview Press.

Gold, R.L. 1974. A Comparative Study of the Impact of Coal Development on the Way of Life of People in the Coal Areas of Eastern Montana and Northeastern Wyoming. Denver, CO: Northern Great Plains Resources Program.

Gramling, R.B. 1980a. Labor Survey of East St. Mary Parish. *In* East St. Mary Parish, Economic Growth and Stabilization Strategies. R.B. Gramling and E.F. Stallings, eds. Baton Rouge, LA: Louisiana Department of Natural Resources, pp. 80–108.

Gramling, R.B. 1980b. Population Growth in East St. Mary Parish Area. *In* East St. Mary Parish, Economic Growth and Stabilization Strategies. R.B. Gramling and E.F. Stallings, eds. Baton Rouge, LA: Louisiana Department of Natural Resources, pp. 65–75.

Gramling, R.B. 1983. Sociological Analysis of Energy Activity on Lafayette, Louisiana. *In* Energy and Economic Growth in Lafayette, LA: 1965–1980. D.P. Manuel, ed. Lafayette, LA: University of Southwestern Louisiana, pp. 167–193.

Gramling, R.B. 1984a. East St. Mary Parish: A Case Study. *In* The Role of Outer Continental Shelf Oil and Gas Activities in the Growth and Modification of Louisiana's Coastal Zone. R.B. Gramling and S. Brabant, eds. Lafayette, LA: U.S. Department of Commerce, National Oceanic and Atmospheric Administration, Louisiana Department of Natural Resources, pp. 171–198.

130

Gramling, R.B. 1984b. Housing in the Coastal Zone Parishes. *In* The Role of Outer Continental Shelf Oil and Gas Activities in the Growth and Modification of Louisiana's coastal zone. R.B. Gramling and S. Brabant, eds. Lafayette, LA: U.S. Department of Commerce, National Oceanic and Atmospheric Administration, Louisiana Department of Natural Resources, pp. 127–134.

Gramling, R.B. 1996. Oil on the Edge: Offshore Development, Conflict, Gridlock. Albany, NY: State University of New York Press.

Gramling, R.B. and S. Brabant, eds. 1984. The Role of Outer Continental Shelf Oil and Gas Activities in the Growth and Modification of Louisiana's Coastal Zone. Lafayette, LA: U.S. Department of Commerce, National Oceanic and Atmospheric Administration, Louisiana Department of Natural Resources.

Gramling, R.B. and S. Brabant. 1986. Boomtowns and Offshore Energy Impact Assessment: The Development of a Comprehensive Model. Sociological Perspectives 29(2):177–201.

Gramling, R.B. and W.R. Freudenburg. 1990a. A Closer Look at Local Control: Communities, Commodities, and the Collapse of the Coast. Rural Sociology 55(4):541–558.

Gramling, R.B. and W.R. Freudenburg. 1990b. Opportunity-Threat, Development, and Adaptation: Toward a Comprehensive Framework for Social Impact Assessment. Rural Sociology 57(2):216–234.

Gramling, R.B. and S.B. Laska. 1993. A Social Science Research Agenda for the Minerals Management Service in the Gulf of Mexico. New Orleans, LA: U.S. Department of the Interior, Minerals Management Service, Gulf of Mexico OCS Region. OCS Study MMS 93-0017.

Gulliford, A. 1989. Boomtown Blues: Colorado Oil Shale, 1885–1985. Niwot, CO: University Press of Colorado.

Heberle, R. 1948. Social Consequences of the Industrialization of Southern Cities. Social Forces 27(October):29–37.

Higgins, R. 1999. Labor Camping: Life in a Louisiana "Bunkhouse." Draft manuscript on file. Prepared by the University of Arizona, Bureau of Applied Research in Anthropology. New Orleans, LA: U.S. Department of the Interior, Minerals Management Service, Gulf of Mexico OCS Region.

Hughes, D.W., J.M. Fannin, W. Keithly, W. Olatubi, and J. Guo. 2002. Lafourche Parish and Port Fourchon, Louisiana: Effects of the Outer Continental Shelf Petroleum Industry on the Economy and Public Service. Part 2. New Orleans, LA: U.S. Department of the

Interior, Minerals Management Service, Gulf of Mexico OCS Region. OCS Study MMS 2001-020.

Ives, B. and W. Schulze. 1976. Boomtown Impacts of Energy Development in the Lake Powell Region. Albuquerque, NM: University of New Mexico, Resource Economics Program.

Jones, L.L., S.H. Murdock, and F.L. Leistritz. 1988. Economic-Demographic Projection Models: An Overview of Recent Developments for Infrastructure Analysis. *In* Local Infrastructure Investment in Rural America. T.G. Johnson, B.J. Deaton, and E. Segarra, eds. Boulder, CO: Westview Press, pp. 87–97.

Keithley, D.C. 2001. Lafourche Parish and Port Fourchon, Louisiana: Effects of the Outer Continental Shelf Petroleum Industry on the Economy and Public Services, Part 1. New Orleans, LA: U.S. Department of the Interior, Minerals Management Service, Gulf of Mexico Region. OCS Study MMS 2001-019.

Kohrs, E.V. 1974. Social Consequences of Boom Town Growth in Wyoming. Paper presented to the Regional Meeting of the American Association for the Advancement of Science, Laramie, Wyoming.

Kuhn, T. 1970. The Structure of Scientific Revolutions. Chicago, IL: University of Chicago Press.

Laska, S.B. ed. 1993. Impact of Offshore Oil Exploration and Production on the Social Institutions of Coastal Louisiana. New Orleans, LA: U.S. Department of the Interior, Minerals Management Service, Gulf of Mexico OCS Region. OCS Study MMS 93-0007.

Leistritz, F.L. 1992. An Economic Perspective on the Theory and Practice of Social Impact Assessment: Economic Focus. *In* Proceedings of the Twelfth Annual Gulf of Mexico Information Transfer Meeting. New Orleans, LA: U.S. Department of the Interior, Minerals Management Service, Gulf of Mexico OCS Region. OCS Study MMS 92-0027. Pp. 211–215.

Leistritz, F.L. and S.H. Murdock. 1981. The Socioeconomic Impact of Resource Development: Methods for Assessment. Boulder, CO: Westview Press.

Lindstedt, D.M., L.L. Nunn, J.C. Holmes, and E.E. Willis. 1991. History of Oil and Gas Development in Coastal Louisiana. Resource Information Series No. 7. Baton Rouge, LA: Louisiana Geological Survey.

Little, R.L. 1977. Some Social Consequences of Boom Towns. North Dakota Law Review 52:401–425.

Louis Berger Group, Inc. 2004. Deepwater Program: OCS-Related Infrastructure in the Gulf of Mexico. New Orleans, LA: U.S. Department of the Interior, Minerals Management Service, Gulf of Mexico OCS Region. OCS Study MMS 2004-027.

Lovejoy, S.B. 1977. Local Perceptions of Energy Development: The Case of the Kaiparowits Plateau. Bulletin No. 62. Los Angeles, CA: Lake Powell Research Project.

Luke, R.T., E.S. Schubers, G. Olsson, and F.L. Leistritz. 2002. Socioeconomic Baseline and Projections of the Impact of an OCS Onshore Base for Selected Florida Panhandle Communities. Volume I: Final Report. New Orleans, LA: U.S. Department of the Interior, Minerals Management Service, Gulf of Mexico OCS Region. OCS Study MMS 2002-024.

Luton, H.H. and D.E. Austin. 2004. Gulf of Mexico Challenges. University of Arizona and the U.S. Department of the Interior, Minerals Management Service.

Manuel, D.P. 1980. East St. Mary Parish in the 1970s: The Economics of a Sustained Energy-Impact. *In* East St. Mary Parish, Economic Growth and Stabilization Strategies. R.B. Gramling and E.F. Stallings, eds. Baton Rouge, LA: Louisiana Department of Natural Resources, pp. 44–48.

Manuel, D.P., ed. 1983. Energy and Economic Growth in Lafayette, LA: 1965–1980. Lafayette, LA: University of Southwestern Louisiana.

Manuel, D.P. 1985. Unemployment and Drilling Activity in Major Energy-Producing States. Journal of Energy and Development 10:45–62.

Manuel, D.P. 1997. The Role of OCS Activity in the Economic Growth of Morgan City. *In* Outer Continental Shelf Impacts, Morgan City, Louisiana. E.F. Stallings and T.F. Reilly, eds. Baton Rouge, LA: Louisiana Department of Transportation and Development, pp. 28–105.

Maruggi, V. and G. Saussy. 1985. Migration in Louisiana, 1970 to 1980: An Indicator of the State Economy's Performance, 2 vols. Research Study 52. New Orleans, LA: University of New Orleans, College of Business Administration, Division of Business and Economic Research, 109 pp.

Maruggi, V. and C.R. Wartenberg. 1996. Louisiana Net Migration, 1980–1990: The Oil Bust Reflected. Research Study No. 60. New Orleans, LA: University of New Orleans, College of Business Administration, Division of Business and Economic Research.

McEvoy, J., III, ed. 1977. The Social Consequences of Environmental Change: A Handbook for Environmental Planning. New York, NY: John Wiley.

McGuire, T.R. and A. Gardner. 2003. Contract Drillers and Causal Histories along the Gulf of Mexico. Human Organization 62(3):218–228.

McKeown, R.L. and A. Lantz. 1977. Rapid Growth and the Impact on Quality of Life in Rural Communities: A Case Study. Glenwood Springs, CO: Colorado West Regional Mental Health Center.

Moen, E.W. 1986. Women: Gemeinschaft in Boomtowns. *In* Differential Social Impacts of Rural Resource Development. P.D. Elkind-Savatsky, ed. Boulder, CO: Westview Press, pp. 161–183.

Murdock, S.H. and F.L. Leistritz. 1979. Energy Development in the Western United States: Impact on Rural Areas. New York, NY: Praeger.

Murdock, S.H., F.L. Leistritz, R.R. Hamm, and S.-S. Hwang. 1984. An Assessment of the Accuracy and Utility of Socioeconomic Impact Assessments. *In* Paradoxes of Western Energy Development: How Can We Maintain the Land and the People If We Develop? C.M. McKell, D.G. Browne, E.C. Cruze, W.R. Freudenberg, R.L. Perrine, and F. Roach, eds. Boulder, CO: Westview Press, pp. 265–296.

National Academy of Sciences. 1978. National Academy of Sciences Findings and Recommendations (Appendix 1). *In* Study Design for Resource Management Decisions: OCS Oil and Gas Development and the Environment. Washington, DC: U.S. Department of the Interior, Bureau of Land Management.

National Research Council. 1992. Assessment of the U.S. Outer Continental Shelf Environmental Studies Program, III: Social and Economic Studies. Washington, DC: National Academy Press.

Olien, R.M. and D.D. Olien. 1982. Oil Booms: Social Change in Five Texas Towns. Lincoln, NE: University of Nebraska Press.

Pikul, R.P. and R. Rabin. 1974. Program Plan for Environmental Effects of Energy. Prepared by the Mitre Corporation and the National Science Foundation. Contract 74-SP-0827 (Project 1700) MTR-6726. Washington, DC: U.S. Department of the Interior, Bureau of Land Management.

Plater, J.R., J.Q. Kelley, W.W. Wade, and R.T. Mott. 2000. Economic Effects of Coastal Alabama and Destin Dome Offshore Natural Gas Exploration, Development, and Production. New Orleans, LA: U.S. Department of the Interior, Minerals Management Service, Gulf of Mexico OCS Region. OCS Study MMS 2000-044.

Plater, J.R. and W.W. Wade. 2001. Forecasting Environmental and Social Externalities Associated with OCS Oil and Gas Development: The Offshore Environmental Cost

Model. Herndon, VA: U.S. Department of the Interior, Minerals Management Service. OCS Study MMS 2001-017.

Porter, B.J. 1992. Socioeconomic Advantages to Outer Continental Shelf Activities in the Gulf of Mexico. *In* Proceedings of the Twelfth Annual Gulf of Mexico Information Transfer Meeting. New Orleans, LA: U.S. Department of the Interior, Minerals Management Service, Gulf of Mexico OCS Region. OCS Study MMS 92-0027. Pp. 66–70.

Priest, T. 2003. History and Evolution of the Offshore Oil and Gas Industry in Southern Louisiana: Exploration technology and the Federal Lease Sales That Opened up the Gulf of Mexico. *In* Proceedings: Twenty-Second Annual Gulf of Mexico Information Transfer Meeting. M. McKay and J. Nides, eds. New Orleans, LA: U.S. Department of the Interior, Minerals Management Service, Gulf of Mexico OCS Region. OCS Study MMS 2003-073. Pp. 288–295.

Ringholz, R.C. 1989. Uranium Frenzy: Boom and Bust on the Colorado Plateau. New York, NY: W.W. Norton.

Schweid, R. 1989. Hot Peppers: Cajuns and Capsicum in New Iberia, Louisiana. Reprint edition. Berkeley, CA: Ten Speed Press.

Scott, L.C. 1978. The Changing Structure of the Louisiana Economy: 1940–1976. Louisiana Business Review (September):2–7,13.

Scott, L.C. 1981. The Louisiana Economy in the 1970s: A Decade of Growth and Internal Shifts. Louisiana Business Review (Spring):8–12.

Seydlitz, R., S. Laska, V. Baxter, S. Brabant, C. Forsyth, R. Seydlitz, and R. Thayer. 1993. Development and Social Problems: The Impact of the offshore Oil Industry on Suicide and Homicide Rates. Rural Sociology 58(1):93–110.

Seydlitz, R. and S.B. Laska, eds. 1994. Social and Economic Impacts of Petroleum "Boom and Bust" Cycles. New Orleans, LA: U.S. Department of the Interior, Minerals Management Service, Gulf of Mexico OCS Region. OCS Study MMS 94-0016.

Shrimpton, M. and K. Storey. 2001. The Effects of Offshore Employment in the Petroleum Industry: A Cross-National Perspective. Herndon, VA: U.S. Department of the Interior, Minerals Management Service. OCS Study MMS 2001-041.

Smith, M.F. 2000. Report on the 1999 Minerals Management Service Social and Economic Studies Conference. Herndon, VA: U.S. Department of the Interior, Minerals Management Service, Environmental Studies Program.

Stein, M.R. 1964. The Eclipse of Community. New York, NY: Harper and Row.

Summers, G.F. and K. Branch. 1982. Human Responses to Energy Development. *In* Energy Resource Communities. G.F. Summers and A. Selvik, eds. Madison, WI: MJM Publishing for the Institute of Industrial Economics, pp. 23–59.

Summers, G.F. and A. Selvik, eds. 1982. Energy Resource Communities. Madison, WI: MJM Publishing for the Institute of Industrial Economics.

Tobin, L.A. 2001. Post-Displacement Employment in a Rural Community: Why Can't Women and Oil Mix? Ph.D. dissertation. Baton Rouge, LA: Department of Sociology, Louisiana State University.

Tolbert, C.M. 1995. Oil and Gas Development and Coastal Income Inequality: A Comparative Analysis. New Orleans, LA: U.S. Department of the Interior, Minerals Management Service, Gulf of Mexico OCS Region. OCS Study MMS 94-005.

Tolbert, C.M. and J. Beggs. 2003. The Coastal Division of Industrial Labor Over Time and Space. Baylor University and Louisiana State University.

Trillin, C. 1979. U.S. Journal: Morgan City, La. New Yorker (January 15):90–93.

Vanclay, F. and D.A. Bronstein, eds. 1995. Environmental and Social Impact Assessment. Chichester, England: John Wiley.

Wade, W.W., J.R. Plater, and J.Q. Kelley. 1999. History of Coastal Alabama Natural Gas Exploration and Development: Final Report. New Orleans, LA: U.S. Department of the Interior, Minerals Management Service, Gulf of Mexico OCS Region. OCS Study MMS 99-0031.

Wallace, B., J. Kirkley, T. McGuire, D. Austin, and D. Goldfield. 2001. Assessment of Historical, Social, and Economic Impacts of OCS Development on Gulf Coast Communities. Volume 2: Narrative Report. New Orleans, LA: U.S. Department of the Interior, Minerals Management Service, Gulf of Mexico OCS Region. OCS Study MMS 2001-027.

Wallace, B., J. Duberg, and J. Kirkley. 2003. Dynamics of the Oil and Gas Industry in the Gulf of Mexico: 1980–2000. New Orleans, LA: U.S. Department of the Interior, Minerals Management Service, Gulf of Mexico OCS Region. OCS Study MMS 2003-004.

Walsh, A.C. 1985. Effects of a Boom-Bust Economy: Women in the Oil Field. Free Inquiry 13(2):133–136.

Weber, B.A. and R.E. Howell, eds. 1982. Coping with Rapid Growth in Rural Communities. Boulder, CO: Westview Press.

Wilkinson, K.P., J.G. Thompson, R.R. Reynolds, Jr., and L.M. Ostresh. 1982. Local Social Disruption and Western Energy Development: A Critical Review (and Response). Pacific Sociological Review 25(3):275–296, 367–376.

Williams, D.C. and K.B. Horn. 1979. Onshore Impacts of Offshore Oil: A User's Guide to Assessment Methods. Washington, DC: U.S. Department of Commerce, National Ocean Survey, Office of Coastal Zone Management, U.S. Office of Policy Analysis.

Wolf, C.P., ed. 1974. Social Impact Assessment. Stroudsburg, PA: Hutchinson, Dowden and Ross.

Notes

---

[1] For information about the Gulf ESP, see:
<http://www.gomr mms.gov/homepg/regulate/environ/studiesprogram.html>
This site contains information on Gulf Region ongoing studies, completed studies from 1993 to the present, and the Annual Studies Plan. It also includes PDF files of all of the more recent study reports.

[2] Environmental justice is the exception that proves the rule. It is a newer, more narrowly defined version of the original NEPA issue of who benefits and who is burdened. While the assessment of the general question of the distribution of benefits remains inconsistent, under Executive Order 12898 (59 FR 7629, 1994) federal agencies are required to identify any disproportionate, negative impacts of their activities on minority or low-income populations. Therefore, environmental justice now appears regularly as a separate category of effects.

[3] The issue of boomtowns lies dormant in coastal Louisiana during the growth of the 1960s and into the 1970s. As offshore activity accelerated in the late 1970s, academics began to debate the usefulness of the classic SIA model in the Gulf. Gramling and Brabant (1986) argued that the local evolution of the offshore industry gave communities time to adjust and that concentrated work schedules and long commutes to work also mitigated demographic effects (cf. Gale 1986 for a rejoinder). Although published later, this academic exchange occurred prior to the oil price bust. In the bust's immediate aftermath, many assessments of oil's community-level socioeconomic impacts drew on this model (e.g., Brabant 1994; Freudenburg 1992; Gramling 1992; Laska 1993; Seydlitz et al. 1993, 1995).

[4] The current trend in the oil-related, labor-intensive fabrication industry toward the importation of guest workers from Mexico and the Caribbean must be viewed similarly. While it was first stimulated by the mini-boom of the 1990s, it is akin to similar trends in the meat packing and nursery industries and is not driven simply by a lack of local labor (Deseran and Tobin 2003; Donato 2004).

[5] Both models assume a larger A generates larger effects at D. However, this relationship is ambiguous for the Gulf because of local variation in industry mix ($c_1, c_2, \dots c_n$) and in community consequences ($d_1, d_2, \dots d_n$). Thus, while the sum of $c_1, c_2, \dots c_n$ equals C and the sum of $d_1, d_2, \dots d_n$ equals D, the concentration of white-collar employment in Houston may mean little to the Gulf industry (B) or to Morgan City (d_1), but much to the commuting suburb of Mandeville, Louisiana (d_2).

[6] For example, consider the always-sensitive issue of race and racism. To show racial discrimination in the oil industry in the 1920s, 1940s, or 1960s is not to prove an effect; rather, the demonstration supports the unsurprising conclusion that this industry reflects imperfections of the society from which it sprang. An "effect" would be a change in racial outcomes. There is some evidence from the 1940s (Jones and Parenton 1951; Brasseaux et al. 1994) and the 1990s (Gardner 2000; Tobin 2001) that job creation by the petroleum industry opened up opportunities for African Americans and other minorities in south Louisiana that did not exist in other rural areas of the state. While this effect seems likely and is predicted by labor-queuing theory, proving it is difficult in the mishmash of history.

137

THE CULTURAL CONTEXT OF OIL DEVELOPMENT IN CALIFORNIA'S COASTAL REGION: CONTRASTING COMMUNITY RESPONSES DURING THE EXUBERANT ERA

MICHAEL R. ADAMSON[1]

Petroleum development in California's coastal region, incorporating the counties of San Luis Obispo, Santa Barbara, and Ventura (see Figure 4), occurred during two cultural eras that conditioned local perceptions of the oil industry and the socioeconomic changes that their activities produced. Cultural exuberance characterized the pre–1965 period of development.[2] Thereafter, oil activity took place during an environmentalist era. Local responses to oil development occurred within these larger cultural contexts. Yet within these cultural eras, community responses differed, depending on their urban growth preferences. This paper presents an overview of the national cultural contexts within which coastal oil development occurred and illustrates how local responses differed during the period of cultural exuberance, using the cases of Santa Barbara and Ventura. In Santa Barbara, oil development was not a "fit" within a qualitative growth strategy. At the same time, Ventura pursued a quantitative growth strategy that welcomed oil development and its contributions to its economy and society, even as oil activity produced short-term social stresses during booms.[3]

The period before 1965, was a time of unprecedented cultural enthusiasm nationally for technology, technological transformation, and modern times. As Thomas P. Hughes has observed, during this period, "Americans created the modern technological nation," and, he argues, "the century of technological enthusiasm [which he locates from 1870–1970] was the most characteristic and impressively achieving century in the nation's history." Technological advances that were cause for national celebration became embedded in systems that included machines, processes, transportation and communications networks, organizations, and people. The culture supported the building of systems that people understood as the foundation of modern life. Few observers imagined any limits on what might be achieved. This cultural exuberance peaked during the interwar period. In the wake of the dropping of the atomic bombs on Hiroshima and Nagasaki, a countercultural movement that cast doubt on the perceived positive impacts of technological invention and improvement began to emerge. But it was only in the 1970s that this movement gained political traction and social momentum, spurred by such events as the Santa Barbara oil spill (Akin 1977; Hart 1998; Hughes 1989; Jordan 1994; Sarewitz 1996).

During the exuberant era, there were few legal restrictions on the development of oil resources in the coastal region, especially in onshore areas where most of the exploration and production occurred. (Most downstream activities occurred in Los Angeles or San Francisco, where operators shipped the crude oil and natural gas that they produced.) Where operators found oil

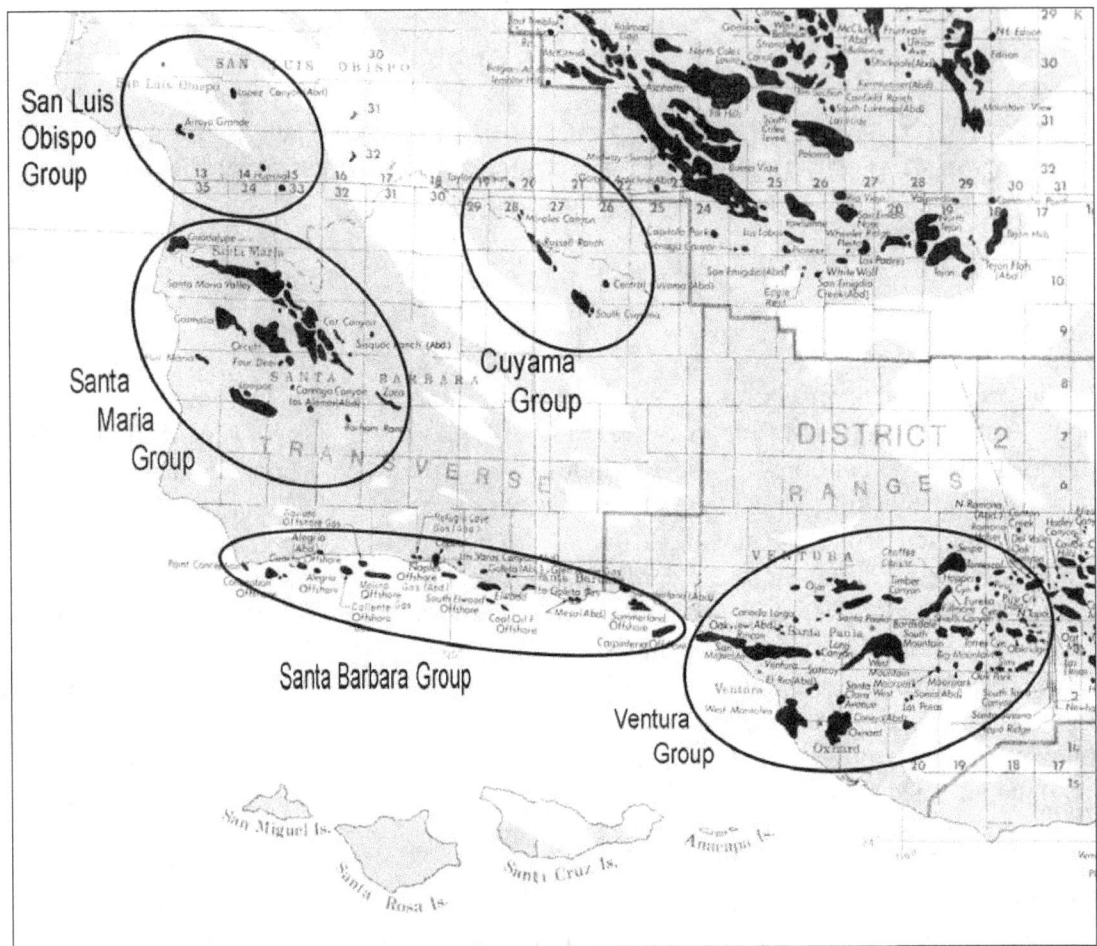

onshore, they generally developed it to the maximum extent possible, given geological and market conditions, technological capacity, managerial expertise, and capital availability (see

Figure 4: County oil and gas fields by group.

Source: California Division of Oil and Gas 1982.

Figures 5 and 6). The tidelands and offshore stories are rather more complicated, involving disputes with the federal government, but prior to 1965, the state of California supported the orderly development of oil and gas reserves under its jurisdiction. In general, the location of reserves, technology, and market conditions determined the character and extent of local oil and gas exploration and production (Schmitt et al. 2003).

By contrast, after 1965 oil activity in the coastal region took place during an environmental era when many people discounted the purported positive impacts of technological invention and improvement, questioned the values of the technological society, and scrutinized its effects on the natural environment. Distrust of technology, and the institutions that deployed it, replaced the enthusiasm that had prevailed (Hughes 1989). The movement gained crucial political and social momentum, especially as it concerned local offshore oil development, after the 1969 Santa Barbara oil spill convinced many coastal residents that ocean, marine, and urban environments were particularly vulnerable to oil development along the California coast.

140

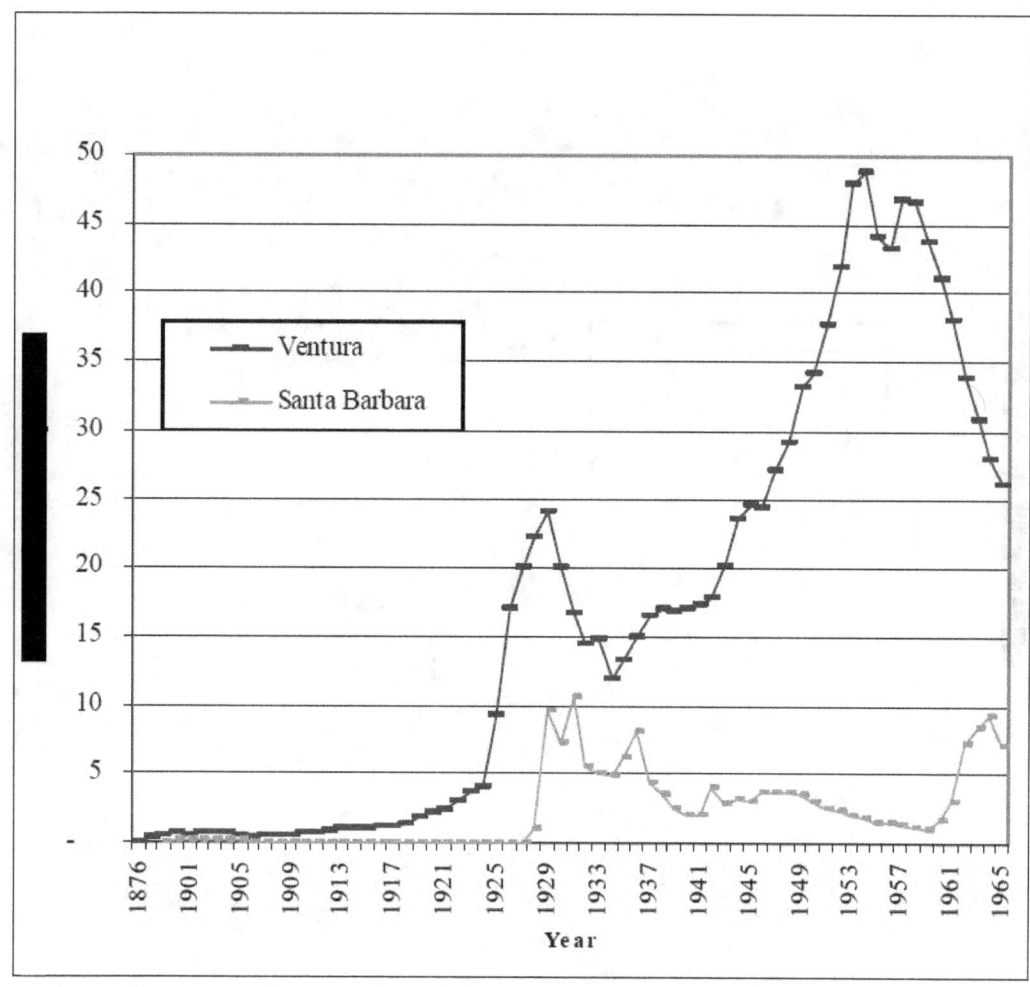

Figure 5: Crude Oil Production: Santa Barbara and Ventura Groups, 1876–1965.

Sources: California Division of Mines and Mining 1927–1929; California Division of Oil and Gas 1929–1966; California State Mining Bureau 1915–1927; Petroleum World 1925.

Community anti-oil backlash to the Santa Barbara oil spill was immediate and sustained, as civic, non-oil-related business, and citizens' organizations called for new restrictions on offshore oil activity, or its termination. New regulations and moratoria on offshore activity raised costs,

delayed development projects that were in process or in the proposal stage, and restricted available locations for new leasing. As a result, the level of offshore oil development in the Santa Barbara channel was much lower than industry executives and federal government officials predicted it would be during the 1960s (Schmitt et al. 2003; Lima 1994; Paulsen et al. 1996; Molotch and Freudenburg 1996; Nevarez et al. 1996; Gramling 1996; Sollen 1998).

During the environmental era, onshore activity was increasingly regulated. However, the impact on actual production during this period was slight, since most development occurred on producing properties that were exempt from many regulations. Operators discovered no

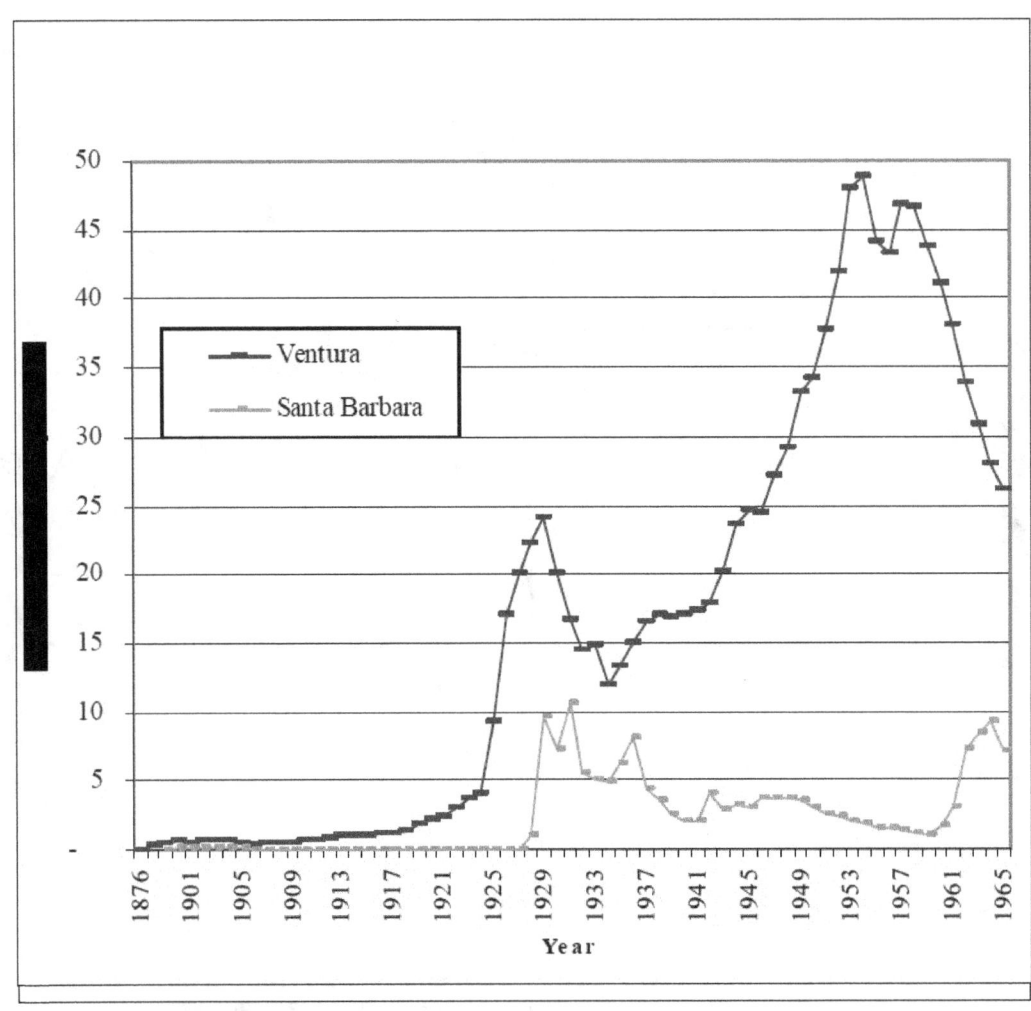

Figure 6: Crude Oil Production: Cuyama Valley and Santa Maria Groups, 1876–1965.

Sources: California Division of Mines and Mining 1927–1929; California Division of Oil and Gas 1929–1966; California State Mining Bureau 1915–1927; Petroleum World 1925.

significant onshore fields in the coastal region after 1965 (Paulsen et al. 1998; Nevarez et al. 1998; Richardson 1962).

Dividing the period of oil activity into two sharply contrasting cultural eras that pre- and postdate 1965 provides a framework for analysis. At the community level, however, perceptions of oil did not change overnight. Areas that developed significant amounts of oil during the exuberant era continued to view oil development favorably well into the environmental era. At the same time,

communities that opposed oil development or experienced little oil activity during the exuberant era opposed oil activity more vigorously during the environmental era. In other words, path dependency has been demonstrated as far as cultural attitudes are concerned, even as all coastal communities increased their opposition to oil over the past thirty years.

142

Based on studies funded by the U.S. Minerals Management Service, Pacific OCS Region, three points may be made regarding the attitudes of coastal communities on the oil industry during the environmental era: (1) broader cultural attitudes supported community efforts to restrict oil activity, both offshore and onshore, (2) coastal communities increasingly adopted measures to slow or stop growth, switching from the pursuit of quantitative or qualitative growth strategies, and (3) regimes regulating oil evolved over a thirty-year period. Thus 1965 may be seen as a demarcation of cultural attitudes rather than as a marker for policy implementation. Perhaps what is notable about the environmental era is the convergence of local attitudes on oil, namely that coastal communities could do without it and that local polities should be proactive in obstructing petroleum exploration and extraction.

Contrasting Local Growth Strategies

Venturans and Santa Barbarans differed in their opinions of oil activity in their communities even as both communities pursued urban growth. For Ventura oil was a "fit" within a perspective that viewed as progress quantitative growth spurred by industrial development. At the same time, Santa Barbarans sought to suppress oil development within their community because it did not "fit" within a qualitative growth strategy that valued historical preservation, small-town values, the coastal and marine environment, and an economy based on tourism, resort living, and "clean" or craft industry. At the same time, Santa Barbarans supported county oil development that did not impinge on urban areas.

The classic boomtown model does not provide a fruitful framework for analyzing the development of coastal region communities with respect to petroleum extraction. Just as Roger and Diana Olien found in their study of five West Texas towns, technological improvements made resource depletion in the coastal region a gradual process (Olien and Olien 1982). Operators developed crude oil reserves in communities that sustained, and sustain, a robust agricultural sector. Economic growth and diversification occurred in conjunction with oil activity, particularly after 1945. The region did not "overadapt" to oil. By all accounts, oil—where it was found in substantial commercial quantities—promoted economic growth and employment on an industrial, rather than extractive, model. Economic adaptation and diversification occurred long before the abandonment or severe curtailment of oil field operations.

Substantial economic growth and diversification also took place independently of the oil industry. Given its location on the Pacific Ocean and its climate, the coastal region has historically attracted tourists and retirees, and, more recently, high-technology industries. A more or less permanent military and defense presence was established during World War II. The area also hosted (and hosts) significant institutions of higher learning. The seemingly inexorable expansion of Los Angeles linked much of Ventura county to its metropolitan area in the post-World War II era at the same time that oil production subsided. Owing to economic diversification and the gradual decline of oil production, the central coast never suffered the busts associated with regions that have been far more dependent on oil and gas extraction. The coastal oil booms of the exuberant era were manageable, and produced positive economic benefits in the major oil districts. As a result, the residents of towns and cities that owed much of their growth to oil activity perceived the industry in a positive light.

Quantitative Growth: Ventura as a "Booming" Oil Town

The city of Ventura experienced a number of oil booms, which cumulatively shaped its urban history on an industrial model that local residents viewed positively. Although it experienced a number of booms during the late nineteenth century, which established important land use precedents for the city's urban development during the twentieth century, this section considers the impact of two booms, one during the late 1920s and the other in the early 1950s, which fueled the community's quantitative growth strategy and established the dominant role of oil in the city's economic and social life.

During the exuberant era, Ventura "encouraged oil to the fullest extent," as the editors of *California Oil World* (1930), a trade journal, observed. Residents embraced an industrial ideal based on disciplined work in oil and gas extraction. Contemporary observers recognized the short-term problems that accompanied the booms. Yet civic and business leaders, as well as residents, valued the economic and urban growth that the booms produced, and compared Ventura's experiences favorably to those of Los Angeles's suburbs.[4]

As of 1920 Ventura remained a small town of 4,156 people. In 1925, it became a veritable boom town with Associated Oil Company's deep-drilling success in the Ventura Avenue field. Technological advances in drilling rigs and muds enabled operators to overcome the high gas pressures and difficult geological structures that had plagued them since the field's discovery in 1913. Promoters and investors from Los Angeles soon sought leases in the district. To their dismay they found that three major oil firms had leased most of the oil bearing lands, which were divided among large ranches. These firms brought in hundreds of workers to work their leases: 450 workers as of February 1925 and 1,200 as of May 1926. Spurred by the relocation of oil workers and their families, Ventura's population increased by 2,000 between February and September 1925 alone. Overall Ventura grew 25% in 1925 and another 27% in 1926. The depression ensured that by 1930 Ventura's population had yet to reach the 25,000 mark predicted by Ralph B. Lloyd, a major leaseholder. Nevertheless, the city's population nearly tripled, to 11,603. Including unincorporated contiguous areas, Ventura's population neared the 19,000 mark (Los Angeles Directory Co. 1929; Ventura County Star 1925c, 1926a, 1926b, 1926c, 1926d; Ventura Free Press 1925a; Wiker 1925).

From 1925–30, Ventura experienced all of the social pressures traditionally associated with oil booms. Real estate prices increased rapidly, the volume of mail swamped the post office, the demand for electrical hook-ups doubled, and many newcomers could not find adequate housing (McPhee 1925; Ventura Free Press 1925c; Ventura County Star 1925a; Wiker 1925). The city tried to keep pace by putting bond issues to build infrastructure before the voters and letting construction contracts to private firms. As Figure 7 shows, building permits soared in response to residential and commercial demand.

144

Figure 7: Building Permits Issued, City of Ventura, California, 1921–1931.

Source: Ventura County Star 1926e, 1929, 1930, 1931.

Aware of the connotation of the term "boom town," one reporter described Ventura as a "booming town" with scant evidence "of the forced, impermanent, uncivilized growth that was apparent in Santa Fe Springs [the gigantic Los Angeles field discovered in 1919] ... It is not the unhealthy growth of a forced hot-house plant, but rather the startling upspring of a well-tended plant which has found a new source of strength in a potent fertilizer" (McPhee 1925). Or, as Associated Oil's resident geologist in Ventura noted: "The growth of the city of Ventura has been greatly augmented by oil field operations, and, unlike some of the southern [California] cities, the work has gone along slowly but steadily and under conditions which have given a substantial foundation for the city's growth" (Hertel 1924).

Most Venturans were pleased with the growth that the boom spurred. As one department store manager put it, many believed that the business expansion produced by oil development would constitute, "the very best kind" kind of growth—"conservative ... and probably permanent in its

value," rather than the "sky-rocket growth that will lead to a slump after a brief period of artificial inflation." The Ventura county district attorney observed: "The oil developments [had] come at a very opportune time for the city," as they compensated for the local slump in agriculture, one of the U.S. economy's "sick" industries of the period (Quoted in McPhee 1925).

The 1920s oil boom began the transformation of Ventura into an industrial city—a source of pride for local residents. Oil activity spurred the creation of new industries, fueled urban growth and improvements in infrastructure, and resulted in an expansion of the city's boundaries. The city's business leaders self-consciously pursued a quantitative growth path that aimed to attract newcomers from southern California and elsewhere, sustaining high demand for housing, oil services, and commercial goods and services until the depression stopped the boom in its tracks (Ventura Free Press 1925b, 1925d; Ventura County Star 1925b, 1925d). Affordable housing along Ventura Avenue replaced the packing houses and farms that had previously occupied the land, and west Ventura, adjacent to the Avenue field, became a working-class neighborhood and center of a burgeoning oil service industry (Reith 1963). Elsewhere, developers and landowners divided the city into subdivisions and began building both single-family homes and apartment buildings. By the early 1930s frenetic building had combined with the breathing space that the depression provided to enable the city's housing stock to alleviate the population pressures of the latter half of the 1920s.

With the boom of the 1950s, spurred primarily by the tapping of deeper zones in existing fields, oil was the leading industry in Ventura and the largest factor shaping its urban development. Residents saw oil companies as key contributors to the city's economic growth and cultural advancement—in a word, progress. Oil workers recognized that the durability of the Ventura district's fields provided stable employment and a settled community (Paulsen et al. 1996). Direct and indirect oil sector employment made the industry a pervasive social and economic influence.[5] Taxes supported schools. (The assessed value of all Ventura county oil properties increased from $23.5 million in 1941 to $164.5 million in 1953, 55% of the county's assessed valuation.) The major operators paid higher-than-average wages and sponsored numerous cultural and educational activities (Paulsen et al. 1996).

A survey of 200 Ventura residents taken at the height of the postwar oil boom, in 1954, found no evidence suggesting that the city was experiencing any of the problems commonly associated with oil booms. More than half of the people surveyed had come to Ventura since 1936: more for business than any other reason. More than half of those surveyed indicated that they preferred Ventura to any other place, and fully 92% of the respondents intended to stay. The study concluded that the "average" Ventura resident was hard-working, earned a relatively high income, was secure economically and content emotionally, placed a high value on the lack of class distinctions among the city's residents, and possessed outstanding community spirit. He or she also cited good schools and numerous parks and opportunities for recreation—all supported by oil money—as key amenities of city life (Reith 1963).

Set in an increasingly urban and economically diversifying area, and sustained over a long period of time, oil development did not mean boom and bust for Ventura in the classic meaning of the terms. Thus the community's residents viewed local oil activity, even during the booms, in positive terms. This was in keeping with national cultural values that equated industrial development and quantitative urban growth with progress.

In Ventura rapid urban expansion created short-term problems in terms of public health and order, housing, education, and government. Yet oil development also offered economic opportunities that compensated for these problems. Oil activity sustained economic development. Those who benefited from it appreciated the wealth-creating effects of oil. To be sure, there were periods of economic stagnation, but these were closely linked to conditions in the national economy, most notably during the depression of the 1930s.

After 1965 urbanization and economic diversification diluted the importance of oil activity to the Ventura district. Increasingly it served as a "bedroom community" for Los Angeles and a home for high-technology and other "clean" businesses. Ventura county as a whole became increasingly professional and middle-class, with the service sector employing one-fourth of the labor force as of 1985. Oil declined in relative and absolute significance. Venturans began to assess ongoing oil activities from the environmentalist perspective, resulting in the adoption of an increasingly restrictive regulatory regime that governed oil activity within the county (Paulsen et al. 1996).

Qualitative Growth: The Santa Barbara Exception

During the exuberant era, Santa Barbara was unique among coastal communities in moving to restrict the development of oil within its urban setting. While most California cities pursued quantitative growth on an industrial model, Santa Barbara sought to grow qualitatively, in keeping with the desires of its well-heeled residents to conserve the natural environment that initially attracted them to this "American Nice." Local civic and business leaders supported them, recognizing that the growth that they desired depended to a large extent on Santa Barbara's beaches, mountains, and climate (Adamson 2004; Starr 1990).

Santa Barbara's attitude toward oil activity was decidedly of the "not-in-my-backward" variety. Local leaders and residents, many of whom were conservationist in outlook, objected to it for aesthetic and socioeconomic reasons. At the same time, many local leaders viewed oil activity elsewhere in Santa Barbara county in positive terms.

Santa Barbarans self-consciously set their community apart from the Los Angeles model embraced by much of California. Santa Barbara was a "relaxed" community, where most residents were interested in arts, culture, leisure, and other activities beyond the "mere routine of making money." It neither "welcomed change" nor followed other cities in "rush[ing] into quick-money schemes and manufacturing plants that desolate the countryside with fumes and smoke." It was "free from unpleasantly distracting and compelling financial ambitions and urges dictated by the transient intellectual fashions of the day." Quantitative-growth minded Los Angeles, by contrast, was overcrowded and smoggy: the quintessential example of big city "hurry and scurry" and where "high-pressure and ulcers and cut-throat way[s] of doing business" were endemic (Santa Barbara News-Press 1949a; Walker 1949). Thus, Santa Barbarans agreed wholeheartedly with Charles Fletcher Lummis, local historian and author, who wrote in 1933 that "the worst curse that could befall Santa Barbara would be the craze of GET BIG! Why big? Run down to Los Angeles for a few days—see that madhouse! You'd hate to live there!" (Quoted in Storke 1958: 286)[6]

At the same time, leading citizens, such as newspaper publishers Reginald G. Fernald and Thomas More Storke, wanted the city to grow and develop in a progressive, non-industrial manner that was compatible with its natural setting. As a 1929 *Santa Barbara Morning Press* editorial made clear, many of the city's political and business leaders promoted the type of planned, qualitative growth that allowed for the development of the waterfront, parks, and a bird sanctuary, the construction of miles of paved streets, the extension of the water system, and large-scale home development. As an advertisement, which appeared in the *Santa Barbara Daily News* on March 15, 1930, proclaimed, boosters and business owners alike sought "a bigger and better Santa Barbara" in keeping with the city's climate, scenic beauty, and other attractions. This approach can be seen in the efforts of the city's Plans and Planting Committee to beautify Santa Barbara, which were closely linked to the efforts of real estate interests to develop posh Hope Ranch and the Riviera.[7]

Through his *Santa Barbara News-Press*, the area's largest newspaper, Storke energetically promoted Santa Barbara as a place congenial to "smokeless," small-scale, and decentralized manufacturing, scientific laboratories, craft-oriented enterprise, and college and university life. Such an economy would also support a broad expansion of the middle-class. Santa Barbara would grow qualitatively, yet "free of unwise crowding and smog and the other unpleasant by-products of growth and wealth that are not well planned to harmonize with nature's plans and offerings" (Storke 1958: 255–8, 269–70). For "the welfare of the area depend[ed] upon its beauty and attractiveness ... natural beauty is a natural resource, to be developed intelligently or lost" (Santa Barbara News-Press 1949e).

Thus, during the 1940s, Storke and his newspaper campaigned relentlessly for the Cachuma Project, a water system that tapped the Santa Ynez River to provide a long-term solution to the city's water needs. The *News-Press* billed Cachuma as "the city's greatest business project," instructing readers to think of water supplies in terms of attracting jobs, rather than simply in terms of keeping gardens green and flowers in bloom (Santa Barbara News-Press 1949b, 1949c). With the voter approval of the project in 1949, the *News-Press* proclaimed: "We are on our way again to a finer as well as a bigger Santa Barbara .... We have what we need for the kind of growth we want—beautiful homes, a beautiful land to live in, life worth living" (Santa Barbara News-Press 1949d).

Until the 1970s Santa Barbara continued to develop in a manner in keeping with the qualitative growth model. Construction of the University of California campus began in 1955. Thereafter electronics, computer, and defense firms set up shop in the area. And, of course, Santa Barbara continued to distinguish itself as a premier tourist and resort location.

During the exuberant era, Santa Barbara was not the no/slow-growth-oriented city that it became during the environmental era. Nevertheless, oil did not fit into the qualitative growth model laid out by Storke and supported by local political and business leaders. Santa Barbara's exuberant-era responses to local oil activity are best understood in the context of a qualitative growth, rather than modern ecological, perspective. Oil activity was potentially detrimental to Santa Barbara's social, economic, and cultural environment. Santa Barbarans objected to it on these grounds. Though they had to tolerate drilling under the contemporary regulatory regime, they sought to limit its extent even as their actions ultimately had little impact on actual oil production prior to 1965.

If oil exploration and production occurred far enough away (that is, in rural Santa Barbara county), however, it was applauded for the taxes and jobs that it generated. Santa Barbara's civic leaders, businessmen, and editors enthused about developments at Elwood, Goleta, and elsewhere along the south Santa Barbara coast. For example, the *Santa Barbara Morning Press* opined that the promising Goleta oil field would be "a wonderful asset to Santa Barbara" if its output and duration met initial expectations. If the field proved to be "a good one," there would be "much additional prosperity in store" for Santa Barbara. Moreover, the field had the advantage of not being "close to residential property or to a section of the seashore that is much used at present by the public" (Santa Barbara Morning Press 1928a). Hugh Martin, a member of the Santa Barbara Lions Club, expressed the sentiments of many Santa Barbarans when he explained the discovery of the gigantic Elwood field in terms of giving the city the best of both worlds. It was "far enough away to avoid damaging our esthetic [*sic*] values," yet it was "close enough to give this city full benefit of the large sums to be spent in [its] development. Every merchant and property owner of the city will benefit" (Quoted in Santa Barbara Morning Press 1928b). For Martin, the economic benefits made oil drilling worth it, as long as he didn't have see it. Possible damage to the coastal and marine environments was apparently not an issue (though, it should be noted, physical waste of crude oil was a major one).

Conclusion

Comparing local responses to oil activity during the exuberant era suggests that there is path dependency in community attitudes, even as national cultural values modify or reinforce previously held views. During the environmental era there has been convergence in opinion among coastal communities regarding oil development. Yet many of the oil-related activities that commenced prior to 1965 have continued. The greatest impact of environmental-era policy has been on new activity. Thus offshore activity has been much more restricted relative to onshore activity in terms of developing the reserves of the extractive region.

References

Adamson, M.R. 2004. The Makings of a Fine Prosperity: Thomas M. Storke, The Santa Barbara News-Press, and the Campaign to Approve the Cachuma Project. Journal of Urban History 30 (January):189–212.

Akin, W.E. 1977. Technocracy and the American Dream: The Technocrat Movement, 1900–1941. Berkeley: University of California Press.

Bertles, B. 1981. Oral history. Ventura County Historical Society, Ventura, CA. 3 April.

California Division of Mines and Mining. 1929–1929. Annual Report of the State Oil and Gas Supervisor. Vols. 13–14. Sacramento, CA: State Printing Office.

California Division of Oil and Gas. 1929–1966. Annual Report of the State Oil and Gas Supervisor. Vols. 15–51. Sacramento, CA: State Printing Office.

California Division of Oil and Gas. 1982. Oil and Gas Prospect Wells Drilled in California Through 1980. Sacramento, CA: State Printing Office.

California Oil World. 1930. Ventura Grows. 8 May.

California State Mining Bureau. 1915–1927. Annual Report of the State Oil and Gas Supervisor. Vols. 1–12. Sacramento, CA: State Printing Office.

Gramling, R. 1996. Oil on the Edge: Offshore Development, Conflict, Gridlock. Albany: State University of New York Press.

Hart, D.M. 1998. Forged Consensus: Science, Technology, and Economic Policy in the United States, 1921–1953. Princeton, NJ: Princeton University Press.

Hertel, F.W. 1924. Ventura's Submarine Pipe Line, The Record 5 (November):10–12.

Hughes, T.P. 1989. American Genesis: A Century of Invention and Technological Enthusiasm, 1870–1970. New York: Viking.

Jordan, J.M. 1994. Machine-Age Ideology: Social Engineering and American Liberalism, 1911–1939. Chapel Hill: University of North Carolina Press.

Lima, J.T. 1994. The Politics of Offshore Energy Development. Ph.D. dissertation. Department of Political Science, University of California, Santa Barbara.

Los Angeles Directory Co. 1929. Ventura County Directory, 1930. Los Angeles, CA: Los Angeles Directory Co.

McPhee, D.G. 1925. Solution of Deep Drilling Problem in the 'Avenue Field' Rejuvenates a Pioneer Oil District. Petroleum World (April):33, 96.

Miranda, R. and D. Rosdil. 1995. From Boosterism to Qualitative Growth: Classifying Economic Development Strategies. Urban Affairs Review 30 (July):868–879.

Molotch, H. and W.R. Freudenburg. 1996. Santa Barbara County: Two Paths. Camarillo, CA: U.S. Department of the Interior, Minerals Management Service, Pacific OCS Region. OCS Study MMS 96-0036.

150

Nevarez, L., H. Molotch, and W.R. Freudenburg. 1996. San Luis Obispo County: A Major Switching. Camarillo, CA: U.S. Department of the Interior, Minerals Management Service, Pacific OCS Region. OCS Study MMS 96-0037.

Nevarez, L., H. Molotch, P. Shapiro, and R. Bergstrom. 1998. Petroleum Extraction in Santa Barbara County, California: An Industrial History. Camarillo, CA: U.S. Department of the Interior, Minerals Management Service, Pacific OCS Region. OCS Study MMS 98-0048.

Olien, R.W. and D.D. Olien. 1982. Oil Booms: Social Change in Five Texas Towns. Lincoln: University of Nebraska Press.

Paulsen, K., H. Molotch, and W.R. Freudenburg. 1996. Ventura County: Oil, Fruit, Commune, and Commute. Camarillo, CA: U.S. Department of the Interior, Minerals Management Service, Pacific OCS Region. OCS Study MMS 96-0035.

Paulsen, K., H. Molotch, P. Shapiro, and R. Bergstrom. 1998. Petroleum Extraction in Ventura County, California: An Industrial History. Camarillo, CA: U.S. Department of the Interior, Minerals Management Service, Pacific OCS Region. OCS Study MMS 98-0047.

Petroleum World. 1925. Crude Oil Production in California from the Industry's Inception to January 1, 1925. (January):70–71.

Reith, G.L. 1963. Ventura: Life Story of a City. Ph.D. dissertation. Department of Geography, Clark University, Worchester, MA.

Richardson, R.L. 1962. California Oil's Past, Present, and Future: Coastal District. Paper presented at the Annual California Regional Meeting of the Society of Petroleum Engineers, Bakersfield, CA (November).

Santa Barbara Morning Press. 1928a. Goleta's Oil Field. 28 July.

Santa Barbara Morning Press. 1928b. Refuse $15,000,000 Goleta Lease. 9 August.

Santa Barbara News-Press. 1949a. An Industry and Activities That Will Add to More Than One Kind of Income. 22 February.

Santa Barbara News-Press. 1949b. This City's Greatest Business Project. 18 May.

Santa Barbara News-Press. 1949c. Water Means Business and Wages as Well and Flowers and Trees. 2 August.

Santa Barbara News-Press. 1949d. We Have Voted Ourselves the Makings of a Great and a Fine Prosperity. 23 November.

Santa Barbara News-Press. 1949e. Planning Has Been Made More Important by Assurance of Water Supply and Growth. 24 November.

Sarewitz, D. 1996. Frontiers of Illusion: Science, Technology, and the Politics of Progress. Philadelphia, PA: Temple University Press.

Schmitt, R.J., J.E. Dugan, and M.R. Adamson. 2003. Industrial Activity and Its Socioeconomic Impacts: Oil and Three Coastal California Counties. Camarillo, CA: U.S. Department of the Interior, Minerals Management Service, Pacific OCS Region. OCS Study MMS 2002-049.

Sollen, R. 1998. An Ocean of Oil. Juneau, AK: Denali Press.

Starr, K. 1990. Material Dreams: Southern California Through the 1920s. New York: Oxford University Press.

Storke, T.M. (in collaboration with W.A. Tompkins). 1958. California Editor. Los Angeles: Westernlore Press.

Ventura County Star. 1925a. Our Greatest Need Today. 13 July.

Ventura County Star. 1925b. More Business. 14 August.

Ventura County Star. 1925c. City Grows 25 Pct. in Year. 31 December.

Ventura County Star. 1925d. Ventura, 1926! 31 December.

Ventura County Star. 1926a. 1926 Will See Record Set for Ventura Oil Development—Lloyd. 26 March.

Ventura County Star. 1926b. Years More of Drilling, Says Lloyd. 29 March.

Ventura County Star. 1926c. Realty Men Tell of This City's Great Opportunity. 2 April.

Ventura County Star. 1926d. Phenomenal Progress Made in All Lines, Figures Show. 31 December.

Ventura County Star. 1926e. The Record for 6 Years. 31 December.

Ventura County Star. 1929. Ventura Shows Large Growth During 1929. 31 December.

Ventura County Star. 1930. Here's How Ventura Developed in 1930. 31 December.

Ventura County Star. 1931. Figures Show Bright Spots in 1931s Record. 31 December.

Ventura Free Press. 1925a. Industries of Ventura Have Pay Rolls Running into Thousands Monthly. 21 January.

Ventura Free Press. 1925b. Here's Way to Aid Ventura to Grow. 10 February.

Ventura Free Press. 1925c. Derth [sic] of Housing Service Now Drives Many to Camp. 11 June.

Ventura Free Press. 1925d. Campaign to Sell Ventura Approved. 23 June.

Viehe, F.W. 1981. Black Gold Suburbs: The Influence of the Extractive Industry on the Suburbanization of Los Angeles, 1890–1930. Journal of Urban History 8 (November):3–26.

Walker, H.L. 1949. Santa Barbara as a Place to Live and Work. Santa Barbara News-Press. 31 August.

Wiker, E.E. 1925. Business Fine in Ventura; Oil Wells, Building Cause Activity. Ventura County Star. 10 September.

Notes

[1] Not to be quoted without the permission of the author.

[2] The coastal region of one of California's three major oil producing regions. With the development of the San Ardo field after World War II, it also included Monterey county. The other two extractive regions were the Los Angeles basin (Los Angeles and Orange counties) and the San Joaquin valley (Fresno, Kern, and Kings counties).

[3] The terms "qualitative growth" and "quantitative growth" may be found in the taxonomy of Miranda and Rosdil (1995). The authors' taxonomy includes quantitative, qualitative, historical preservation, environmentally harmful, and redistributive growth.

[4] On the petroleum-based industrial ideal, see Viehe (1981).

[5] According to the 1960 California Statistical Abstract, 2,800 oil, gas, and mineral extraction workers resided in Ventura County. At the time of the region's only major strike in the petroleum industry of the twentieth century, in 1948, some 1,500 workers belonged to local 120 of the Oil Workers Union (Bertles 1981).

[6] In the early 1900s, Lummis promoted Hispanic imagery as the best development metaphor for Los Angeles. The rapid pace of the latter's growth prompted him to relocate to Santa Barbara. His *Stand Fast, Santa Barbara!*, which the *Santa Barbara Morning Press* published in 1922, was reprinted in pamphlet form and widely distributed locally (Starr 1990).

[7] My thanks to W. Elliot Brownlee for this point.

The Department of the Interior Mission

As the Nation's principal conservation agency, the Department of the Interior has responsibility for most of our nationally owned public lands and natural resources. This includes fostering sound use of our land and water resources; protecting our fish, wildlife, and biological diversity; preserving the environmental and cultural values of our national parks and historical places; and providing for the enjoyment of life through outdoor recreation. The Department assesses our energy and mineral resources and works to ensure that their development is in the best interests of all our people by encouraging stewardship and citizen participation in their care. The Department also has a major responsibility for American Indian reservation communities and for people who live in island territories under U.S. administration.

The Minerals Management Service Mission

As a bureau of the Department of the Interior, the Minerals Management Service's (MMS) primary responsibilities are to manage the mineral resources located on the Nation's Outer Continental Shelf (OCS), collect revenue from the Federal OCS and onshore Federal and Indian lands, and distribute those revenues.

Moreover, in working to meet its responsibilities, the **Offshore Minerals Management Program** administers the OCS competitive leasing program and oversees the safe and environmentally sound exploration and production of our Nation's offshore natural gas, oil and other mineral resources. The MMS **Minerals Revenue Management** meets its responsibilities by ensuring the efficient, timely and accurate collection and disbursement of revenue from mineral leasing and production due to Indian tribes and allottees, States and the U.S. Treasury.

The MMS strives to fulfill its responsibilities through the general guiding principles of: (1) being responsive to the public's concerns and interests by maintaining a dialogue with all potentially affected parties and (2) carrying out its programs with an emphasis on working to enhance the quality of life for all Americans by lending MMS assistance and expertise to economic development and environmental protection.

www.ingramcontent.com/pod-product-compliance
Lightning Source LLC
Chambersburg PA
CBHW080810180526
45168CB00006B/2394